ALAN TURING AND HIS CONTEMPORARIES

BCS THE CHARTERED INSTITUTE FOR IT

Our mission as BCS, The Chartered Institute for IT, is to enable the information society. We promote wider social and economic progress through the advancement of information technology science and practice. We bring together industry, academics, practitioners and government to share knowledge, promote new thinking, inform the design of new curricula, shape public policy and inform the public.

Our vision is to be a world-class organisation for IT. Our 70,000-strong membership includes practitioners, businesses, academics and students in the UK and internationally. We deliver a range of professional development tools for practitioners and employees. A leading IT qualification body, we offer a range of widely recognised qualifications.

Further Information
BCS, The Chartered Institute for IT,
First Floor, Block D,
North Star House, North Star Avenue,
Swindon, SN2 1FA, United Kingdom.
T +44 (0) 1793 417 424
F +44 (0) 1793 417 444
www.bcs.org/contactus

ALAN TURING AND HIS CONTEMPORARIES

Building the world's first computers

Simon Lavington (editor)

Published by British Informatics Society Limited (BISL), a wholly owned subsidiary of BCS The Chartered Institute for IT First Floor, Block D, North Star House, North Star Avenue, Swindon, SN2 1FA, UK. www.bcs.org

ISBN: 978-1-90612-490-8
PDF ISBN: 978-1-78017-105-0
ePUB ISBN: 978-1-78017-106-7
Kindle ISBN: 978-1-78017-107-4

British Cataloguing in Publication Data.
A CIP catalogue record for this book is available at the British Library.

Disclaimer:
The views expressed in this book are of the author(s) and do not necessarily reflect the views of BCS or BISL except where explicitly stated as such. Although every care has been taken by the authors and BISL in the preparation of the publication, no warranty is given by the authors or BISL as publisher as to the accuracy or completeness of the information contained within it and neither the authors nor BISL shall be responsible or liable for any loss or damage whatsoever arising by virtue of such information or any instructions or advice contained within this publication or by any of the aforementioned.

Typeset by Lapiz Digital Services, Chennai, India.
Printed at CPI Antony Rowe Ltd, Chippenham, UK.

CONTENTS

Contents

AUTHORS

Christopher P Burton MSc, FIET, FBCS, CEng graduated in Electrical Engineering at the University of Birmingham. He worked on computer hardware, software and systems developments in Ferranti Ltd and then ICT and ICL, nearly always being based in the Manchester area, from 1957 until his retirement from the industry in 1989. He is a member of the Computer Conservation Society (CCS) and led the team that built a replica of the Manchester Small-Scale Experimental Machine (SSEM). Other roles in the CCS have included chairmanship of the Elliott 401 Project Group and of the Pegasus Project Group, and more recently investigating the feasibility of building a replica of the Cambridge EDSAC. For replicating the SSEM he was awarded an honorary degree by the University of Manchester, the first Lovelace Gold Medal by BCS, The Chartered Institute for IT, and a Chairman's Gold Award for Excellence by ICL.

Martin Campbell-Kelly is Emeritus Professor in the Department of Computer Science at the University of Warwick, where he specialises in the history of computing. His books include *Computer: A History of the Information Machine*, co-authored with William Aspray, *From Airline Reservations to Sonic the Hedgehog: A History of the Software Industry*, and *ICL: A Business and Technical History*. He is editor of *The Collected Works of Charles Babbage*. Professor Campbell-Kelly is a Fellow of BCS, The Chartered Institute for IT, visiting professor at Portsmouth University, and a columnist for the *Communications of the ACM*. He is a member of the ACM History Committee, a council member of the British Society for the History of Mathematics, and a committee member of the BCS Computer Conservation Society. He is a member of the editorial boards of the *IEEE Annals of the History of Computing*, the *International Journal for the History of Engineering and Technology* and the *Rutherford Journal*, and editor-in-chief of the Springer Series in the History of Computing.

Roger Johnson is a Fellow of Birkbeck College, University of London, and Emeritus Reader in Computer Science. He has a BSc in Pure Mathematics and Statistics from the University College of Wales, Aberystwyth and a PhD in Computer Science from London University. He has researched and written extensively on a range of issues concerning the management of large databases. He worked previously at the University of Greenwich and at a leading UK software house. He was Chairman of the BCS Computer Conservation Society from 2003 to 2007 and has served on its committee since its founding. He has lectured and written about the history of computing, notably on the work of the UK pioneer, Andrew D Booth. He also co-authored the first academic paper on the history of the ready reckoner. He has been active in BCS, The Chartered Institute for IT for many years, serving as President in 1992–3 and holding a number of other senior offices. He has represented BCS for many years on international committees, becoming President of the Council of European Professional Informatics Societies (CEPIS) from 1997 to 1999. During his service with CEPIS he was closely involved in establishing the European Computer Driving Licence and the ECDL Foundation. He served as Honorary Secretary of the International Federation for Information Processing (IFIP) from 1999 to 2010. He is currently Chairman of IFIP's International Professional Practice Programme (IP3) promoting professionalism in IT worldwide.

Simon Lavington MSc, PhD, FIET, FBCS, CEng is Emeritus Professor of Computer Science at the University of Essex. He graduated in Electrical Engineering from Manchester University in 1962, where he remained as part of the Atlas and MU5 high-performance computer design teams until he moved to lead a systems architecture group at the University of Essex in 1986. From 1993 to 1998 he also coordinated an EPSRC specially promoted programme of research into Architectures for Integrated Knowledge-based Systems. Amongst his many publications are four books on computer history: *History of Manchester Computers* (1975), *Early British Computers* (1980), *The Pegasus Story: a history of a vintage British computer* (2000); and *Moving Targets: Elliott-Automation and the dawn of the computer age in Britain, 1947–67* (2011). He retired in 2002 and is a committee member of the Computer Conservation Society.

ACKNOWLEDGEMENTS

The Computer Conservation Society is a member group of BCS, The Chartered Institute for IT. The authors wish to acknowledge the financial support of BCS in producing this book and the assistance of Matthew Flynn and the BCS Publications Department. We are grateful to Kevin Murrell, Secretary of the Computer Conservation Society, for arranging the photographs and credits.

Picture credits
Archant, Norfolk: P. 56 (top)

Birmingham Museums Collection Centre: P. 66

Bletchley Park Trust: P. 2 (bottom)

Computer Laboratory, University of Cambridge: Pp. 21; 22 (top and bottom); 24; 27; 28; 29; 30 (top and bottom)

Crown Copyright, with the kind permission, Director GCHQ: P. 2 (top)

From author's private collection (RJ): Pp. 60; 63; 64; 67

From author's private collection (MC-K): P. 54 (bottom)

Elliott-Automation's successors (BAE Systems and Telent plc): Pp. 48 (top, middle and bottom); 51 (top); 54 (top); 56 (bottom); 77

Fujitsu plc.: P. 19 (top)

IBM plc: P. 36 (bottom)

LEO Society: P. 26

Medical Research Council, Laboratory of Molecular Biology: P. 31

National Physical Laboratory: P. 18 (top and bottom)

Royal Society: P. 80

School of Computer Science, University of Manchester: Pp. 34; 36 (top); 39 (top and bottom); 41; 42; 43; 44; 46; 51 (bottom)

Science Museum, London: Front cover; P. 62

St John's College, University of Cambridge: Pp. 3; 5

Twickenham Museum: P. 6

PREFACE

The years 1945–55 saw the emergence of a radically new kind of device: the high-speed stored-program digital computer. Secret wartime projects in areas such as code-breaking, radar and ballistics had produced a wealth of ideas and technologies that kick-started this first decade of the Information Age. The brilliant mathematician and code-breaker Alan Turing was just one of several British pioneers whose prototype machines led the way.

Turning theory into practice proved tricky, but by 1948 five UK research groups had begun to build practical stored-program computers. This book tells the story of the people and projects that flourished during the post-war period at a time when, in spite of economic austerity and gloom, British ingenuity came up with some notable successes. By 1955 the computers produced by companies such as Ferranti, English Electric, Elliott Brothers and the British Tabulating Machine Co. had begun to appear in the marketplace. The Information Age had arrived.

To mark the centenary of Alan Turing's birth, the Computer Conservation Society has sponsored this book to celebrate the efforts of the people who produced the world's first stored-program computer (1948), the first fully functional computing service (1950), the first application to business data processing (1951) and the first delivery of a production machine to a customer (1951). Our book is a tribute not only to stars such as Tom Kilburn, Alan Turing and Maurice Wilkes but to the many other scientists and engineers who made significant contributions to the whole story.

Chapter 1 sets the background to these events, explaining how, and where, the basic ideas originated. Chapters 2–6 describe how teams at five UK locations then built a number of prototype computers based on these ideas. Chapter 7 explains how these prototypes were re-engineered for the market place, leading to end-user applications in science, industry and commerce. The relative influence of Alan Turing in all of this, through his contributions both to the theory and the practice of computing, is summarised in Chapter 8. The book concludes with a technical appendix that gives the specifications and comparative performance of the principal computers introduced in the main text.

Simon Lavington
25 September 2011
lavis@essex.ac.uk

1
THE IDEAS MEN

Simon Lavington

SCIENCE AT WAR

The momentous events of the Second World War saw countless acts
of bravery and sacrifice on the part of those caught up in the conflict.
Rather less perilously, large numbers of mathematicians, scientists and
engineers found themselves drafted to government research establishments where they worked on secret projects that also contributed to
the Allied war effort. This book is about the people who took the ideas
and challenges of wartime research and applied them to the new and
exciting field of electronic digital computer design. It is a complex story,
since the modern computer did not spring from the efforts of one single inventor or one single laboratory. In this chapter we give an overall sense of the people involved and the places in Britain and America
where, by 1945, ideas for new forms of computing were beginning to
emerge.

In Britain the secret wartime establishment that is now the most
famous was the Government Code and Cipher School at **Bletchley Park**
in Buckinghamshire. Bletchley Park together with its present-day successor organisation, the Government Communications Headquarters
(GCHQ), may be well known now but in the 1940s – and indeed right
up to the 1970s – very few people were aware of the code-breaking
activity that had gone on there during the war. The mathematician
Alan Turing was perhaps the most brilliant of the team of very clever
people recruited to work there. In the spirit of the time, let us keep the
story of Bletchley Park hidden for the moment. We shall return to it
after introducing examples of other scientific work that went on in Britain and America during the war.

In both countries research into radar featured prominently. The challenge was to improve the accuracy and range of detection of targets,
for which vacuum tube (formerly called 'thermionic valve') technology

Bletchley Park and Colossus This country mansion in Buckinghamshire was taken over by the Government Code and Cipher School (GCCS) in 1938 and was soon to become the centre for top-secret code-breaking during the war. When activity there was at its height the mansion and numerous temporary outbuildings housed a staff of about 9,000, of whom 80 per cent were women.

Up to 4,000 German messages that had been encrypted by Enigma machines were being deciphered every day. Bletchley Park developed electromechanical machines called Bombes to help decode Enigma messages. From mid 1942 the Germans introduced the formidable Lorenz 5-bit teleprinter encryption machine for High Command messages.

To analyse and decipher the Lorenz messages, mathematicians at Bletchley Park and engineers from the Post Office's Research Station at Dollis Hill developed the Colossus series of high-speed electronic digital machines. Operational from December 1943, these Colossus machines were of crucial importance to the Allied war effort. However,

their design had little impact upon early general-purpose computers for two reasons: firstly, their very existence was not made public until the 1970s; secondly, they were special-purpose machines with very little internal storage.

You can visit Bletchley Park today and see working replicas of a Bombe and a Colossus.

Professor Douglas Hartree
is shown here in about 1935
operating a Brunsviga mechanical
desk calculating machine. Hartree
(1897–1958) was a mathematical
physicist who specialised in
numerical computation and
organised computing resources
during the Second World War.
After the war he took the lead in
encouraging the design and use
of the new prototype universal
stored-program computers for
science and engineering.

and electronic pulse techniques were stretched to the limit. The Telecommunications Research Establishment (TRE) at Malvern, Worcestershire, became a world-class centre for electronics excellence, especially as applied to airborne radar. Research for ship-borne naval radar was carried out at the Admiralty Signals Establishment (ASE) at Haslemere and Witley in Surrey.

In 1945, as hostilities ended, senior people from the various British and American research establishments visited each other's organisations and exchanged ideas. Amongst the subjects often discussed was the task of carrying out the many kinds of calculations and simulations necessary for weapons development and the production of military hardware. During the war scientific calculations had been done on a range of digital and analogue machines, both large and small. The great majority of these calculators were mechanical or electromechanical. In Britain the mathematician and physicist **Douglas Hartree** had masterminded many of the more important wartime computations required by government research establishments. In America one particular research group had decided to overcome the shortcomings of the slow electromechanical calculators by introducing high-speed electronic techniques. It was thus that in 1945, in Pennsylvania, the age of electronic digital computing was dawning.

THE MOORE SCHOOL: THE CRADLE OF ELECTRONIC COMPUTING

A huge electronic calculator called **ENIAC** (Electronic Numerical Integrator and Computer) was developed under a US government contract at the Moore School of Electrical Engineering at the University of Pennsylvania. The spur for ENIAC had been the need to speed up the process of preparing ballistic firing tables for artillery. Leading the development team were two academics: the electrical engineer Presper Eckert and the physicist John Mauchley. As the work of building the huge machine progressed a renowned mathematician from Princeton University, John von Neumann, was also drawn into the project. Von Neumann subsequently

ENIAC Construction of ENIAC (Electronic Numerical
Integrator and Computer) started in secret in 1943 at
the University of Pennsylvania. It was first demonstrated
to the public in February 1946. ENIAC was a magnificent
beast. It contained 17,468 vacuum tubes, 7,200
semiconductor diodes and 1,500 relays, weighed nearly
30 tons and consumed 150 kW of power. It could carry
out 5,000 simple additions or 385 multiplications per
second – a speed improvement of about a thousand
times on the existing mechanical methods.
Plug-boards were used for setting up a problem. The
ENIAC could be programmed to perform complex
sequences of operations, which could include loops,
branches and subroutines, but the task of taking
a problem and mapping it on to the machine was
complex and usually took weeks. Although primarily
designed to compute ballistics tables for artillery,
ENIAC could be applied to a wide range of practical
computational tasks. It was not, however, a universal
stored-program machine that we would now recognise
as truly general purpose.

(in about 1948) used ENIAC for calculations associated with the development of the hydrogen bomb.

Even before ENIAC itself had been completed the team working on it was producing ideas for a successor computer, to be called EDVAC, the Electronic Discrete Variable Automatic Computer. The team's ideas addressed a challenge: how to make ENIAC more general purpose, so that its benefits could be more easily applied to a much wider range of computational tasks. The ideas were written up by John von Neumann in June 1945 in a 101-page document entitled *First draft of a report on the EDVAC*. By 1946 copies of this report were being distributed widely and were read with interest on both sides of the Atlantic. A project to build EDVAC was launched in 1946, but due to organisational problems the machine did not become operational until 1951.

Most importantly, however, the *EDVAC Report* of 1945 contained the first widely available account of what we would now recognise as a general-purpose stored-program electronic digital computer. EDVAC has become formally known as a 'stored-program' computer because a single memory was used to store both the program instructions and the numbers on which the program operated. The stored-program concept is the basis of almost all computers today. Machines that conform to the EDVAC pattern are also sometimes called 'von Neumann' computers, to acknowledge the influence of the report's author.

The June 1945 EDVAC document was in fact a paper study, more or less complete in principle but lacking engineering detail. Once hostilities in the Pacific had ceased there was an understandable desire to consolidate the Moore School's wartime ideas and to explain the details to a wider American audience. Accordingly, the US government funded an eight-week course of lectures in July–August 1946 on the 'Theory and Techniques for Design of Electronic Digital Computers'. Twenty-eight scientists and engineers were invited to attend. Amongst these were just three Englishmen: David Rees, Maurice Wilkes and Douglas Hartree. David Rees had worked at Bletchley Park and then, when the war ended, had joined the Mathematics Department at Manchester University. Maurice Wilkes had worked at TRE during the war and had returned to Cambridge University to resume his leading role at the Mathematical Laboratory (later to become the Computer Laboratory). Douglas Hartree, at that time Professor of Physics at Manchester University but soon to move to Cambridge, was invited to give a lecture on 'Solution of problems in applied mathematics'.

The *EDVAC Report* and the Moore School lectures were the inspiration for several groups worldwide to consider designing their own general-purpose electronic computers. Certainly Maurice Wilkes's pioneering computer design activity at Cambridge University, described in Chapter 3, grew out of the Moore School ideas. The Moore School's activities were also of considerable interest to Rees's Head of Department at Manchester University, **Professor Max Newman**, who had been at Bletchley Park during the war. What happened at Manchester after 1946 is explained in Chapter 4.

Although the ideas promoted by the Moore School were of equal interest to Alan Turing, they were to produce a different kind of effect upon his thinking.

THE UNIVERSAL TURING MACHINE

Alan Turing was a most remarkable man. A great original, quite unmoved by authority, convention or bureaucracy, he turned his fertile mind to many subjects during his tragically short life. Though classed in the Scientific Hall of Fame as a mathematician and logician, he explored areas as diverse as artificial intelligence (AI) and morphogenesis (the growth and form of living things).

Professor Max Newman (1897–1984) was a Cambridge mathematician who joined Bletchley Park in 1942 to work on cryptanalysis. He specified the logical design of the Colossus code-cracking machine. In 1945 Newman moved to Manchester University, where he encouraged the start of a computer design project and promoted its use for investigating logical problems in mathematics.

Alan Turing   This photograph shows Alan Turing in 1946, the year in which he was appointed OBE (Order of the British Empire) for his wartime code-breaking efforts at Bletchley Park. By 1946 he was working at the National Physical Laboratory (NPL) on the design of the ACE computer. Turing's involvement with computers is explained in more detail in Chapter 2 and Appendix B. Here is a summary of his brief but extraordinary life.

1912 Born at Paddington, London, on 23 June
1926–31 Sherborne School, Dorset
1931–4 Mathematics undergraduate at King's College, Cambridge University
1934–5 Research student studying quantum mechanics, probability and logic
1935 Elected Fellow of King's College, Cambridge
1936–7 Publishes seminal paper 'On Computable Numbers', with the idea of the Universal Turing Machine
1936–8 Princeton University – PhD in logic, algebra and number theory, supervised by Alonzo Church
1938–9 Returns to Cambridge; then joins Bletchley Park in September 1939
1939–40 Specifies the Bombe, a machine for Enigma decryption
1939–42 Makes key contributions to the breaking of U-boat Enigma messages
1943–5 A principal cryptanalysis consultant; electronic work at Hanslope Park on speech encryption
1945 Joins National Physical Laboratory, London; works on the ACE computer design
1946 Appointed OBE for war services
1948 Joins Manchester University in October; works on early programming systems
1950 Suggests the Turing Test for machine intelligence
1951 Elected Fellow of the Royal Society; works on the non-linear theory of biological growth (morphogenesis)
1953–4 Unfinished work in biology and physics
1954 Death (suicide) by cyanide poisoning on 7 June

Why was the young Alan Turing, just back from completing a doctorate in America, one of the first mathematicians to be recruited to help with code-cracking at Bletchley Park in 1939? The answer probably lies in a theoretical paper that he had written back in 1935–6, whilst a post-graduate at King's College, Cambridge.

Turing's paper was called 'On Computable Numbers, with an application to the *Entscheidungsproblem*'. In plain English, it was Turing's attempt to tackle one of the important philosophical and logical problems of the time: Is mathematics decidable? This question had been posed by scholars who were interested in finding out what could, and what could not, be proved by a given mathematical theory. In order to reason about this so-called *Entscheidungsproblem*, Turing had the idea of using a conceptual automatic calculating device. The 'device' was a step-by-step process – more a thought-experiment, really – that manipulated symbols according to a small list of very basic instructions.

The working storage and the input–output medium for the process was imagined to be an infinitely long paper tape that could be moved backwards and forwards past a sensing device.

It is now tempting to see Turing's mechanical process as a simple description of a modern computer. Whilst that is partly true, Turing's Universal Machine was much more than this: it was a logical tool for proving the decidability, or undecidability, of mathematical problems. As such, Turing's Universal Machine continues to be used as a conceptual reference by theoretical computer scientists to this day. Certainly it embodies the idea of a *stored program*, making it clear that instructions are just a type of data and can be stored and manipulated in the same way. (If all this seems confusing, don't worry! It is not crucial to an understanding of the rest of this book.)

In the light of his theoretical work and his interest in ciphers, Alan Turing was sent to Bletchley Park on 4 September 1939. He was immediately put to work cracking the German Naval Enigma codes. He succeeded. It has been said that as Bletchley Park grew in size and importance Turing's great contribution was to encourage the other code-breakers in the teams to think in terms of probabilities and the quantification of weight of evidence. Because of this and other insights, Turing quickly became the person to whom all the other Bletchley Park mathematicians turned when they encountered a particularly tricky decryption problem.

On the strength of his earlier theoretical work Alan Turing was recruited by the National Physical Laboratory (NPL) at Teddington in October 1945, as described in Chapter 2. Senior staff at NPL had heard about ENIAC and EDVAC and wished to build a general-purpose digital computer of their own. Turing, they felt, was the man for the job. It is very likely that at NPL Turing saw an opportunity to devise a physical embodiment of the theoretical principles first described in his 'On Computable Numbers' paper. Although he was well aware of the developments at the Moore School and knew John von Neumann personally, Turing was not usually inclined to follow anyone else's plans. Within three months he had sketched out the complete design for his own general-purpose stored-program computer – which, however, did adopt the notation and terminology used in the *EDVAC Report*. For reasons described in Chapter 2, Turing's paper design for what was called ACE, the Automatic Computing Engine, remained a paper design for some years.

PRACTICAL PROBLEMS, 1945–7

To some extent the problems that beset Turing at NPL also dogged other pioneering computer design groups in the immediate post-war years. The main problem was computer storage. Central to the idea of a universal automatic computer was the assumption that a suitable storage system or 'memory' could be built. The *EDVAC Report* was very clear about this, stating that the implementation of a general-purpose computer depended 'most critically' on the engineers being able to devise a suitable store.

Many ideas for storage were tried by the engineers of the time; few proved reliable and cost-effective. The trials and tribulations of the principal early British computer design groups are recounted in Chapters 2 to 6. These groups were in the end successful, and indeed in a couple of cases they outpaced the contemporary American groups in building working computers. It is tempting to believe that progress was helped by a continuation of the spirit of inventiveness that the designers had experienced during their wartime service in government research establishments.

All of the designers of early computers were entering unknown territory. They were struggling to build practical devices based on a novel abstract principle – a *universal computing machine*. It is no wonder that different groups came up with machines of different shapes and sizes, having different architectures and instruction sets and often being rather less than user-friendly.

THE RICH TAPESTRY OF PROJECTS, 1948–54

To set the scene for the rest of this book, the diagram opposite gives a picture of the many **British computer projects** that bridged the gap between wartime know-how and the marketplace. At the top of the diagram we can imagine the people and ideas flowing out of government secret establishments in 1945. At the bottom are the practical production computers that were available commercially in the UK by 1955. In between the arrows show how ideas and technologies fed through universities and research centres into industry and then out into the marketplace. The left-hand box shows that, at the same time, there were a number of classified government projects that remained secret. Surprisingly, Alan Turing's own attempt at practical computer design at NPL, the Pilot ACE, did not bear fruit until 1950.

Of course, Britain was not the only country actively working on high-speed electronic digital computers in the late 1940s. There were at least a dozen pioneering projects in America. Amongst the earliest of

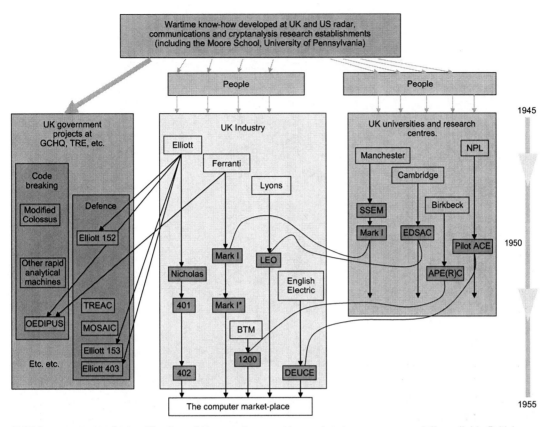

British computer projects The flow of ideas and techniques that came from government wartime research via pioneering prototype projects and into the marketplace as commercially available British computers is shown here. The projects mentioned in the diagram are described in detail in Chapters 2 to 6.

these to become operational were machines called SEAC (May 1950), SWAC (August 1950), ERA 1101 (December 1950), UNIVAC (March 1951), WHIRLWIND (March 1951), IAS (summer 1951) and EDVAC (late 1951). In Germany Konrad Zuse designed a series of ingenious electromechanical computers between 1938 and 1945, but these were *sequence-controlled* and not *stored-program* machines. In Australia the CSIRAC electronic stored-program computer first worked in November 1949. Its designer, Trevor Pearcey, had graduated in Physics from Imperial College, London University in 1940 and spent the rest of the war working on radar at the Air Defence Experimental Establishment (ADEE). He moved to Australia in late 1945.

In the next chapter we continue the story of Alan Turing's progression from Bletchley Park to NPL and from thence to Manchester. This represents but one strand of post-war British computing activity. Many other people, as we have already seen, began to be involved in the late 1940s at various places and at various times. It is an intriguing tale.

2
ACES AND DEUCES

Simon Lavington

TURING'S FIRST COMPUTER DESIGN

After almost three years of intensive day-to-day code-breaking activity at Bletchley Park, Alan Turing was gradually moved towards longer-term planning. From 7 November 1942 to 23 March 1943 he was part of a British Joint Staff Mission visiting the United States, where, amongst other things, he saw a secure speech cipher system at Bell Labs. This intrigued him, and he believed he could improve on the design. From the autumn of 1943 Turing was spending two days a week working on speech encipherment at Hanslope Park, which was about ten miles north of Bletchley Park and was the home of various secret communications projects. By the autumn of 1944 Turing was working full time at Hanslope Park on the speech project, which was by now known as Delilah. This activity gave Turing some first-hand experience of electronic design – including some primitive experiments with a form of storage called a 'delay line'. The prototype Delilah began to work in the summer of 1945.

Meanwhile, unrelated developments had been taking place at the National Physical Laboratory (NPL) at Teddington. In September 1944 the mathematician John Womersley had become head of a new Mathematics Division at NPL. One of his briefs was to oversee the development of electronic devices for rapid scientific computing. In the spring of 1945 Womersley went on a two-month tour of American computing installations and became the first non-American to be allowed access to ENIAC.

In June 1945 Womersley met Alan Turing, to whom he showed the draft *EDVAC Report*. Womersley had read 'On Computable Numbers', and he persuaded Turing to take a job as Senior Scientific Officer in the NPL Mathematics Division, starting on 1 October 1945. Turing was charged with designing an electronic universal computing machine. This was undoubtedly a subject on which he had already been pondering. There is also little doubt that the project was seen at the time by NPL as Britain's answer to the EDVAC proposals.

By the end of 1945, and in the remarkably short time of three months, Alan Turing had finished his first NPL report. It was entitled *Proposed electronic calculator*. Historians now judge it to be the first substantially complete description of a practical stored-program computer. The typewritten document was very detailed, running to the modern equivalent of 83 printed pages including 25 pages of diagrams. It was what we would now call a register-level and system-level description rather than a precise engineering design, though it did contain sample electronic circuits, an estimate of the cost (£11,200, equivalent to perhaps £250,000 in 2012) and a guess that the computer could be built within about a year. (These estimates soon proved to be wildly optimistic.) There was an 11-page section giving a detailed mathematical analysis of delay-line storage.

Alan Turing's 1945 report makes reference to John von Neumann's *EDVAC Report* and indeed uses the same basic notation and terminology. It is therefore interesting to compare the two. In the light of hindsight, we now judge von Neumann's report to be less complete and less general purpose, placing more emphasis on a computer as a numerical calculator intended for scientific applications. In contrast, Turing's report described a more complex and more flexible machine, indicating a much wider range of applications. Turing firmly believed that it was desirable for one program to be able to modify another. He demonstrated a better understanding of nested subroutine calling and return, and his report contains much practical discussion about program preparation.

Turing's report was also quite different from the *EDVAC Report* in three matters of detail. Firstly, and this is something that may seem strange to modern eyes, Turing's machine did not have what we would recognise as a main accumulator. Secondly, there seemed to be no recognisable conditional transfer (branch, or jump) instruction. Thirdly, Turing required a programmer to specify the address of the next instruction to be obeyed, instead of the default being that instructions followed each other sequentially. It is not easy to explain these points until we have revealed more about the approach of other pioneers at the Universities of Cambridge and Manchester, and elsewhere, in designing their own computers. A technical explanation of the unique features of Turing's report is therefore postponed to Chapter 8. A detailed comparison of the characteristics of six early British computers is given in Appendix A.

Womersley gave Turing's proposed computer the name ACE: Automatic Computing Engine. The word 'engine' was a deliberate reference to Charles Babbage's unfinished Analytical Engine of a hundred years before. Womersley did not, it seems, anticipate the scale of the staff and resources that would be needed to implement ACE. In Womersley's defence, Turing was not an easy person to work with, and, throughout 1946, he was continually modifying his ACE design. Indeed, before long he was writing to a friend:

> In working on the ACE I am more interested in the possibility of producing models of the brain than in the practical applications to computing.

TOIL AND TROUBLE

In June 1946 NPL reached an agreement with engineers at the Post Office Research Station at Dollis Hill, designers of the Colossus code-cracking machines, that Dollis Hill would develop **mercury d elay-line storage** for ACE. In the event, Dollis Hill was overburdened with repairing bomb-damaged telephone exchanges, and the agreement was terminated in March 1947.

In May 1946 NPL had recruited Jim Wilkinson to work half time and Mike Woodger to work full time helping Alan Turing with the mathematical aspects of the ACE project. ACE then went through several modifications and many programs were desk tested, but little effort was put into electronic design. Sir Charles Darwin, the boss of NPL, sought in turn the collaboration of TRE, Cambridge and Manchester with the ACE project, but all three groups became too busy implementing their own computer designs.

Then, in January 1947, Harry Huskey, an American ex-ENIAC engineer, arrived to spend a year's attachment to NPL, at the suggestion of Douglas Hartree. Huskey set about designing a simplified version of the ACE, called the Test Assembly, but this work was stopped by Darwin in September of that year. Huskey's comment was that 'morale in the Mathematics Division has collapsed'. However, on the positive side, NPL at last recruited two engineers with relevant wartime experience of pulse electronics, Ted Newman and David Clayden, to work on ACE. At the same time two more mathematicians, Gerald Alway and Donald Davies, joined the team.

At this point – September 1947 – something dramatic happened. The team leader, Alan Turing, decided to ask for leave of absence and took himself off for a year's sabbatical at Cambridge University.

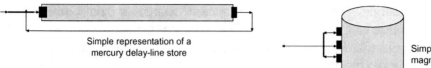

Simple representation of a
mercury delay-line store

Simple representation of a
magnetic drum store

Delays and drums: early storage technologies One form of early computer storage depended upon the great difference in speed between electronic pulses and sound waves. Electronic pulses representing binary ones and zeros can be converted into pulses of sound, best thought of as acoustic shock waves, by piezoelectric crystal transducers. In the 1940s the sound waves were often transmitted along a metal tube containing mercury, to be reconverted into electronic pulses by a receiving crystal at the remote end. Each sound wave took about one millisecond (one thousandth of a second) to travel along a tube of mercury about 5 feet (1.5 m) long. If electronic pulses were produced every one microsecond (a millionth of a second) inside the main computer, then about one thousand such pulses, when converted to sound, could be 'stored' as they travelled slowly along the tube of mercury. In Turing's ACE proposal, 1,024 pulses representing 1,024 binary digits were stored in each of several *mercury delay lines*.

There were at least three problems with mercury delay lines. They were expensive per stored bit, they were sensitive to changes in temperature, and the bits could only be accessed sequentially (i.e., one after another). There was a need for a cheaper and more robust storage technology. One way of providing this was via a *magnetic drum store*.

Electronic pulses can be made to record sequences of binary digits on a magnetic surface. The problem is how to read back these digits at high speeds. If a spinning disk or drum is coated with magnetic material, recording and reading heads can be placed close to the spinning surface, and binary information can be 'written to' and 'read from' the surface. This is similar to the technology used in a modern computer's hard drive. The total storage capacity of a drum depends on many factors but mainly on the dimensions of the drum and the number of individual tracks of information arranged round the periphery. Early drum stores were relatively ponderous pieces of equipment, but they did provide economical storage. Once again, however, the bits could only be accessed sequentially.

Much ingenuity was exercised by Turing and other computer pioneers in overcoming the essentially sequential access properties of delay lines and drums, in an attempt (as we now see) to obtain the effect of so-called *random-access* properties, as given by modern random-access memory (RAM).

What induced Turing to leave? One can only guess that, in addition to becoming disillusioned with the lack of progress on hardware construction, his fertile mind was racing ahead to consider ever more challenging uses for universal digital computers. Let us pause in the story of ACE to explain what was preoccupying him.

INTELLIGENCE AND ARTIFICIAL INTELLIGENCE

For several years prior to arriving at NPL Alan Turing had been musing about the possibilities of intelligent machines. It is not surprising that when listing future applications of ACE in his 1945 report he included the possibility of checking for winning moves in a game of chess. Turing wrote:

> Can the machine play chess? It could fairly easily be made to
> play a rather bad game. It would be bad because chess requires
> intelligence. We stated at the beginning of this section that
> the machine should be treated as entirely without intelligence.

There are indications however that it is possible to make the machine display intelligence at the risk of its making occasional serious mistakes. By following up this aspect the machine could probably be made to play very good chess.

During the autumn of 1946, when news of ENIAC and the ACE plans became public knowledge, newspapers started publishing articles that spoke of the computer as an 'electronic brain'. NPL staff tried to calm expectations, but this did not stop the *Daily Telegraph* reporting on 7 November 1946 that

Dr Turing, who conceived the idea of ACE, said he foresaw the time, possibly in 30 years, when it would be as easy to ask the machine a question as to ask a man.

Today we have become used to search engines such as Google whizzing around the internet, guessing at answers to our half-baked factual questions. In 1946 Turing was considering a much more challenging form of intellectual debate with a machine.

Turing developed his own notion of *artificial intelligence* further in a lecture to the London Mathematical Society in February 1947. He said:

Let us suppose that we have set up a machine with certain initial instruction tables [i.e. programs], so constructed that these tables might on occasion, if good reason arose, modify those tables. One can imagine that after the machine has been operating for some time, the instructions would have altered out of all recognition, but nevertheless still be such that one would have to admit that the machine was still doing very worthwhile calculations ... When this happens I feel one is obliged to regard the machine as showing intelligence.

During his sabbatical at Cambridge Turing became more interested in thinking processes and mechanised learning and renewed his interest in game theory. Away from the day-to-day anxieties about the tangled ACE project he was able to do some serious thinking about future possibilities. By August 1948 he had completed a lengthy report for NPL entitled *Intelligent Machinery*. This was partly speculative but, interestingly, included a detailed technical section on the properties of neural networks. All this went down like a lead balloon with NPL! Sir Charles Darwin, its head, judged the report as 'not suitable for publication', and it was filed away. An edited version was eventually published posthumously in 1969, by which time artificial intelligence was becoming a popular topic for research.

After moving to Manchester University in October 1948 Turing continued his thoughts about mechanised learning, though as a background activity. Then in July 1949 the Ratio Club, a very influential gathering of psychologists, physiologists, mathematicians and engineers, was formed in London to discuss issues in cybernetics. Turing soon became a member and went to meetings every few months. The Ratio Club became a forum for him to discuss his ideas of machine intelligence over the next few years.

It was at Manchester in 1949 that there took place one of the earliest serious debates on artificial intelligence. On 27 October a formal discussion on 'The Mind and the Computing Machine' was held in the Philosophy Department at Manchester University. Besides Turing, this meeting was attended by many eminent UK academics, amongst them Max Newman, Michael Polanyi and J Z Young. As a result of it, Turing wrote up his views as a 27-page paper entitled 'Computing machinery and intelligence', which appeared in the philosophical journal *Mind* in 1950. In trying to answer the question 'Can machines think?' Turing devised the well-known **Turing Test**.

Another indication of Turing's preoccupations at this time is that he stated in his paper that 'the nervous system is certainly not a discrete-state machine' but went on to estimate the storage capacity of the human brain in terms of binary digits.

By 1950 Turing's interests had apparently turned away from artificial intelligence and towards morphogenesis, the growth and form of living things. However, he wrote that morphogenesis 'is not altogether unconnected with' his interest in brain cells and the physiological basis of memory and pattern recognition. Alas, Turing was not to publish any more specific papers on artificial intelligence. Nevertheless, today many people would describe him as the 'father of AI'.

The Turing Test   In his 1950 paper Turing posed the question: 'Can machines think?'. He chose to discuss this question in terms of an 'imitation game', which is now usually referred to as the Turing Test. His original game was played by three people, a man (A), a woman (B) and an interrogator (C) who may be of either gender. To quote Turing:

> The interrogator stays in a room apart from the other two. The object of the game for the interrogator is to determine which of the other two is the man and which is the woman. He knows them by labels X and Y, and at the end of the game he says either 'X is A and Y is B' or 'X is B and Y is A'. The interrogator is allowed to put questions to A and B. In order that tones of voice may not help the interrogator the answers should be written, or better still, typewritten. The ideal arrangement is to have a teleprinter communicating between the two rooms.

Turing asked what will happen when a machine takes the part of A in this game. He suggested that if the responses from the computer are indistinguishable from those of a human, then the computer can be said to be thinking.

PILOT ACE ARRIVES AT LAST

Back at NPL in September 1947 Ted Newman and David Clayden had started serious electronics work on what was now being called the ACE Pilot Model. With Alan Turing away on sabbatical, Jim Wilkinson took charge of the Pilot ACE developments. Turing handed in his resignation to the NPL Director on 28 May 1948. After completing his *Intelligent Machinery* report in August he went on a well-earned holiday to Switzerland, followed by some time in the Lake District and then in Wales. He finally arrived at Manchester University shortly after 2 October to take up the position of Deputy Director of the Royal Society Computing Machine Laboratory.

In April 1948 a separate Electronics Section was established at NPL, with F M Colebrook in charge. This was the turning point: Woodger, Wilkinson, Alway and Davies were temporarily moved to the NPL Electronics Section to join Newman and Clayden. The Pilot ACE project, now loosely based on Harry Huskey's 1947 Test Assembly version of Turing's ideas, forged ahead. Construction began early in 1949. The English Electric Co. Ltd, whose chairman, Sir George Nelson, was a member of the NPL Executive Committee, provided a small group to help with the development. The intention was to pave the way for an eventual commercial exploitation of the ACE design.

The **Pilot ACE** first ran a program on 10 May 1950. In comparison with other early British machines, it was compact and fast. It contained about 1,000 vacuum tubes. Initially **Pilot ACE's storage system** consisted of eight, and finally 11, long delay lines of 32 words each, together with eight short (single-word) delay lines called 'temporary stores' (TS). The machine had a theoretical maximum speed of 16,000 instructions per second, though a typical average figure was 5,000 instructions per second. In any case, Pilot ACE was faster than other contemporary British computers by about a factor of five, whilst employing about one-third of the electronic equipment. Technical comparisons are given in more detail in Appendix A.

A **magnetic drum store** was added to the Pilot ACE in 1954.

The Pilot ACE computer at the National Physical Laboratory in 1950. Three of the design team are shown (left to right): G G Alway, E A Newman and J H Wilkinson. 'DSIR' stood for Department of Scientific and Industrial Research, the government body responsible for funding the NPL. Much of the Pilot ACE can be seen today at the Science Museum in London.

The Pilot ACE's delay-line storage system is clearly visible in this photograph of the computer, taken at the Mathematics Division of NPL in 1952. The short delay lines are on the central stand. Two experimental long delay lines are shown in the right foreground. The large box behind the computer's main frame is the temperature-controlled enclosure for the other long delay lines.

**An English Electric DEUCE
computer** at the NPL in 1956. The
pieces of equipment in the left and
right foreground are respectively a
card reader, used for input, and a
card punch, for output.

DEUCE AND OTHERS

The English Electric company, which manufactured
everything from electric trains to jet aircraft, and from
radar to domestic appliances, had loaned personnel to
NPL to help with the construction of the Pilot ACE. The
company then took the design, made a number of minor
improvements, and in 1955 produced a commercially
available version called **DEUCE**. Thirty-three of these
computers were built, of which 12 remained within
English Electric where they were put to work on a range
of engineering problems and computing bureau activity.
In this they benefited from the numerical algorithms and
software already developed by the Mathematics Division at NPL. DEUCE had a primary store consisting of
12 long delay lines, backed by a drum of capacity 8,000
words. There were also a number of shorter lines: two
lines holding four words, three lines holding two words
and four temporary stores holding single words. Appendix A gives more information on DEUCE.

Alan Turing's ACE design had a number of other
descendants. All of them had in common Turing's philosophy of instruction set design, particularly with
provision for the address of the next instruction to be
included in the current instruction. This arrangement
allowed programmers to place each instruction and its
data in optimal positions in store and, by specifying

The descendants of ACE   The family tree of
computers directly influenced by Alan Turing's 1945
report to NPL is shown in this diagram.

The MOSAIC (Ministry of Supply Automatic
Integrator and Computer) was built by the Post
Office for the Radar Research and Development
Establishment, for use in radar signal analysis. It was
huge, containing 6,000 vacuum tubes and three-
quarters of a ton of mercury. The Bendix G15 and the
Packard-Bell PB250 were small American production
computers. The EMI EBM (Electronic Business
Machine) was a small one-off development for the
British Motor Corporation, produced by Electric and
Musical Industries Ltd (EMI). The ACE was NPL's own
one-off implementation of Turing's original proposal,
but with an increased word length and revised
instruction format. It used mercury delay lines, which
by 1958 were becoming obsolescent.

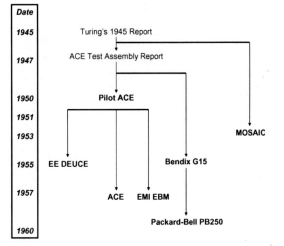

various timing parameters, to maximise the rate at which instructions were obeyed. This led to the term 'optimum programming', also called 'minimum latency coding'. The drawback was that obtaining high computing speeds demanded high programming skills. Once computers had emerged from the laboratory and were applied to a wide range of industrial and commercial problems, the lack of adequately skilled programmers became a big issue. In the words of the Ferranti Sales Director, speaking in 1955, 'Optimum programming was to be avoided because it tended to become a time-wasting intellectual hobby of programmers.'

By the end of the 1950s the underlying technical justification for optimum programming had been entirely removed. This was because bit-serial storage devices such as delay lines had become obsolete, being replaced by storage systems whose access time was independent of the position, or address, of a bit – the so-called *random-access* systems – and most computers now used random-access core stores. The ferrite core store was a welcome upgrade for the earlier, but less reliable, random-access Williams–Kilburn CRT (cathode ray tube) store, a device that will be explained in Chapter 4.

Although Turing's 1945 ACE design was the first substantially complete description of a practical computer, it did not set the pattern for most future machines. Apart from the optimum programming issue described above, there are other factors that made its influence less than might have been expected. It also arrived a year or more after pioneers at a number of other places in England and America had made good progress with their own designs for general-purpose stored-program computers. There is a great deal more computer history to relate before we can properly set Turing's work in context.

3
IVORY TOWERS AND TEA ROOMS

Martin Campbell-Kelly

MAURICE WILKES AND THE CAMBRIDGE UNIVERSITY MATHEMATICAL LABORATORY

Cambridge University had established a computing facility before the Second World War, and there was an indirect link with Manchester University. At Manchester Douglas Hartree, Professor of Applied Mathematics, had learned of the *differential analyser*, an analogue computing machine for solving differential equations that had been invented by the engineer Vannevar Bush at the Massachusetts Institute of Technology in 1930. Hartree built a small **differential analyser out of Meccano** in 1934 and it proved to be surprisingly accurate and effective. So much so, that the theoretical chemist **Professor John Lennard-Jones** at Cambridge University decided to have one made, too. **Maurice Wilkes** – who was then a research student in the Cavendish Laboratory – was an enthusiastic user of the machine. Wilkes was

The Meccano differential analyser at Cambridge University in about 1935, with Maurice Wilkes standing at the right of the picture

later to become a key figure not only in Cambridge but in computing circles worldwide. Indeed, he has sometimes been called the 'father of British computing'.

Computing was growing in importance in the sciences in the 1930s, and in 1937 Lennard-Jones persuaded Cambridge University to establish a Mathematical Laboratory to provide computing facilities and advice for the whole university. He was appointed part-time director of the new laboratory, and Wilkes became the full-time assistant director. In September 1939, however, before the laboratory could really get going, Britain declared war on Germany. The laboratory was taken over by the Ministry of Supply, and Wilkes joined the scientific war effort, for which he worked on radar and operations research. This background in electronics and mathematics, and the contacts he made, would prove very useful after the war when it came to building an electronic computer.

Professor Sir John Lennard-Jones FRS, the founding director of the Mathematical Laboratory

POST-WAR RECONSTRUCTION AND THE STORED-PROGRAM COMPUTER

In October 1945 Wilkes returned to Cambridge University to take full charge of the Mathematical Laboratory. He had two tasks: first, to conduct research into

Maurice Wilkes   Professor Sir Maurice Wilkes FRS (1913–2010) led the Cambridge University Mathematical Laboratory (later called the Computer Laboratory) for over 40 years. A mathematics graduate with an early interest in amateur radio, Wilkes completed a PhD in the propagation of radio waves in 1936. During the war he worked on radar at the Telecommunications Research Establishment (TRE) and on operational research. Back at Cambridge he spearheaded the design of the world's first practical stored-program computer, the EDSAC, and developed the laboratory's academic and research programme. He played an important public role in helping to establish the British Computer Society, serving as its inaugural President from 1957 to 1960. Today his name is honoured in many ways. For example, the Wilkes Award is given annually for the best paper published in a volume of BCS's *Computer Journal*.

The picture shows Wilkes in 1948, kneeling besides a battery of 16 mercury delay lines for the EDSAC computer. The lines, or tubes, were kept in a thermostatically controlled 'coffin' to keep their temperature stable. The 16 tubes in the 'coffin' stored 512 short words, each of which could hold an instruction or a number. EDSAC eventually had 32 delay lines, giving a total storage capacity of 1024 words.

computing machinery and methods; second, to provide a computing service – re-equipping the laboratory with the best available computing facilities and helping scientists to make use of them.

In May 1946 Wilkes had a visit from L J Comrie, who was advising him on re-equipping the laboratory. Comrie was one of Britain's foremost computing experts – he had established the world's first for-profit computing service in London in the 1930s and had prospered during the war. He brought with him a copy of the famous *EDVAC Report*, which had been written by John von Neumann on behalf of the computer group at the Moore School of Electrical Engineering at the University of Pennsylvania. The Moore School had recently completed the ENIAC computing machine, and the EDVAC proposal was a carefully considered design that came out of the ENIAC experience.

There were no photocopiers in those days, so Wilkes stayed up late into the night reading the *EDVAC Report*. He recognised it at once as 'the real thing' and decided that the laboratory had to have an EDVAC-type *stored-program* computer. A few weeks later he received a telegram from the dean of the Moore School. They were organising a summer school in computer design, and would he like to attend? Wilkes would, and did. Unfortunately, because of shipping delays he was only able to attend the latter part of the course, but that was all he needed – he now had a detailed insight into the EDVAC design.

Returning to England on the *Queen Mary* in September 1946, Wilkes began the design of the EDSAC – Electronic Delay Storage Automatic Calculator. The name 'EDSAC' was deliberately chosen to echo that of the EDVAC, so that there should be no doubt about the machine's provenance.

A MEMORY FOR EDSAC

The biggest problem facing all of the computer pioneers was that of building a memory capable of storing at least a thousand instructions and numbers. In 1946 no one had yet done this, anywhere. At the Moore School it had been decided to base the EDVAC on a mercury delay-line memory, so that was what Wilkes also decided to do.

In October 1946 Wilkes had a stroke of luck when he met a newly arrived research student at the Cavendish Laboratory by the name of Tommy Gold. Gold had worked on radar research for the Admiralty during the war and had actually constructed a working delay-line memory for radar echo cancellation. He was able to give Wilkes the necessary constructional data and Wilkes followed his instructions to the letter.

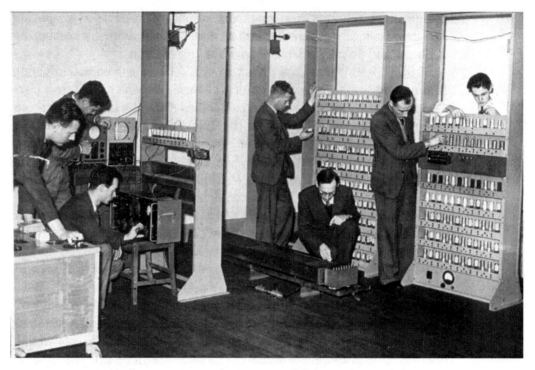

The EDSAC team in 1948 (left to right): G J Stevens, J Bennett, S A Barton, P Farmer, Maurice Wilkes (kneeling), Bill Renwick and R Piggott

Before he could start building the machine Wilkes also needed to recruit an **EDSAC engineering team**. Here Gold came to his aid again. He was able to recommend a seasoned electronics engineer he knew from the Admiralty Signal Establishment, Bill Renwick. Wilkes and Renwick divided responsibility for the EDSAC until it was completed. More technical staff were recruited in the following months.

EDSAC, ACE AND LEO

Funding was not too much of an issue when it came to building the EDSAC. The laboratory was well supported by the university and did not need any external sources of finance. Nevertheless, the costs of the EDSAC were unknown, and, like any academic entrepreneur, Wilkes was alert to funding opportunities. One possibility came from the National Physical Laboratory (NPL).

Towards the end of the war a Mathematics Division had been established in the NPL. This was to have a similar function to that of the Cambridge University Mathematical Laboratory – conducting research and providing computing facilities. However, it would be on a national scale, and its services would be available to both industry and universities. The head of the new Mathematics Division was the mathematician Dr John Womersley, and Alan Turing was leading the electronic computer section.

Womersley wanted to know if it might be feasible to have a joint computer development between Cambridge and the NPL. This idea probably came at the suggestion of Douglas Hartree, who was on the executive committee of the NPL. In December 1946 Wilkes sent his technical proposal for the EDSAC to Womersley. Womersley in turn sought the advice of his electronic computer expert, Turing. Turing was thoroughly dismissive of the EDSAC. He wrote:

> The 'code' which he [Wilkes] suggests is however very contrary
> to the line of development here, and much more in the American
> tradition of solving one's difficulties by means of much equipment
> rather than by thought. I should imagine that to put his code
> (which is advertised as 'reduced to the simplest possible form') into
> effect would require a very much more complex control circuit than
> is proposed in our full-size machine … It is clearly rank folly to
> develop a complex control merely for the sake of a pilot model.

Wilkes did not see Turing's abrasive report until many years later, but Womersley gave him the gist. Cambridge University and the NPL decided to go their separate ways.

Today we can see that there was merit on both sides. Wilkes wanted to keep things simple – which meant following the American lead, using the EDVAC design, and pressing ahead to complete a machine as soon as possible. For him the important thing was to provide a computing facility and to learn about computer programming, not to get bogged down in the details of design. He also felt that Turing's design for the ACE computer, which strove to overcome the slow nature of sequential stores such as serial delay lines, would be short lived because sooner or later true random-access memories would come along, making the ACE design obsolete.

Turing's ACE design, on the other hand, looked to be far more cost-effective than the EDSAC. For the same amount of equipment it would provide very much more computing power. Even the small Pilot ACE version that was eventually built provided more computing power than the EDSAC did, with only a quarter as much equipment. But Wilkes was proved right in the long run.

Meanwhile, the catering and bakery firm of J Lyons & Company was taking an interest in the activities at Cambridge. The firm was well known for its high street cafes, known as 'tea shops'. It was also well known as a pace setter in office mechanisation, and after the war its office and methods experts visited the USA to study developments

LEO The photograph shows LEO, Britain's first computer designed for business, which was based on the EDSAC.

Lyons Electronic Office (or LEO) was the first computer in the world especially designed for commercial work as opposed to scientific computations. Although the central processor was based on EDSAC, the complete LEO was twice the size because of the extra equipment needed for handling large volumes of data. The machine went into service in 1951 and aroused a great deal of interest from other companies, who asked whether it would be possible to have one. As a result, Lyons created a subsidiary, LEO Computers Limited, in 1955 to manufacture office computers. The first machine, LEO II, was an enhanced production version of the original LEO that was four times as fast. Eleven copies of LEO II were sold between 1957 and 1961. LEO III and its successors sold in much greater numbers. During the 1960s LEO Computers was absorbed into ICL (International Computers Ltd), Britain's 'national champion' computer manufacturer. We shall meet ICL again in Chapter 7.

in office automation and electronic computers. They visited Princeton University (where John von Neumann was based), and discovered that the organisation that was furthest ahead with computer development was practically on their own doorstep – Cambridge University!

When they got back to the UK the Lyons team visited the Mathematical Laboratory and agreed with Wilkes that, when it had been completed, they could make a copy of the EDSAC for commercial use. Their computer would be called the **LEO** – the Lyons Electronic Office. In exchange they loaned one of their technical staff to the EDSAC team and provided a grant of about £3,000.

NOT JUST EDSAC

Building EDSAC was only one aspect of the Mathematical Laboratory's work. Wilkes also needed to build up an academic programme and a research community, and provide a computing service. In November 1947 he established a series of fortnightly colloquia – seminars that were attended by representatives from most of the computing groups in the UK. These provided a focus for information exchange in this fast-moving field. A course in numerical methods was established – with Douglas Hartree, who had moved from Manchester in 1946 to become a professor of Physics at Cambridge University, giving the lectures. The first research students joined in 1947–8: John Bennett, David Wheeler and Stanley Gill. They all went on to become major figures in the world of computing.

The **EDSAC sprang into life** on Friday 6 May 1949. It contained some 3,000 vacuum tubes, its input–output was via paper tape, and it performed about 650 operations a second. More technical details are given in Appendix A. The historic first program was a **table of the squares of the integers**. As soon as EDSAC was running the laboratory contacted Lyons, and LEO got the go-ahead.

Why was the machine completed so quickly – ahead of any American one? It was because Wilkes decided on a straightforward design with conservative electronics. For example, the machine operated with a pulse rate of 500 kHz when, if he had been more adventurous, he might have tried to make it twice as fast. But Wilkes reasoned that the EDSAC would be a thousand times faster than anything previously available, and that users would be well enough pleased with the speed and would value having a computer sooner rather than later. Towards the end of June 1949 the laboratory held a conference to celebrate the completion of the machine. It was attended by most of the UK and European computer community – there were 144 delegates.

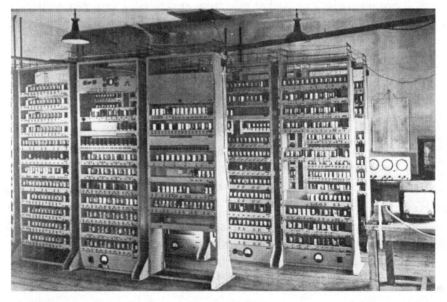

The EDSAC shortly after its completion in May 1949

CAMBRIDGE
EDSAC.
FIRST ACHIEVEMENT
MAY 7ᵗʰ 1949.

0000	0001	0004	0009	0016	0025	0036	0049	0064	0081
0100	0121	0144	0169	0196	0225	0256	0289	0324	0361
0400	0441	0484	0529	0576	0625	0676	0729	0784	0841
0900	0961	1024	1089	1156	1225	1296	1369	1444	1521
1600	1681	1764	1849	1936	2025	2116	2209	2304	2401
2500	2601	2704	2809	2916	3025	3136	3249	3364	3481
3600	3721	3844	3969	4096	4225	4356	4489	4624	4761
4900	5041	5184	5329	5476	5625	5776	5929	6084	6241
6400	6561	6724	6889	7056	7225	7396	7569	7744	7921
8100	8281	8464	8649	8836	9025	9216	9409	9604	9801

FIRST STEPS IN PROGRAMMING

After the conference Wilkes wrote his first real application program, one to integrate Airy's differential equation. This was the kind of problem that arose in the physics of the atmosphere, which Wilkes had studied as a research student in the Cavendish Laboratory. It was actually quite a short program of about 125 instructions, but he got 20 of them wrong! It took him at least a dozen attempts before he finally coaxed the correct results out of the program. He had discovered debugging. He later recalled:

By June 1949 people had begun to realize that it was not so easy to get a program right as had at one time appeared. I well remember when this realization first came on me with full force. The EDSAC was on the top floor of the building and the tape punching and editing equipment one floor below on a gallery that ran round the room in which the differential analyzer was installed. I was trying to get working my first non trivial program, which was one for the numerical integration of Airy's differential equation. It was on one of my journeys between the EDSAC room and the punching equipment that, 'hesitating at the angles of stairs', the realization came over me with full force that a good part of the remainder of my life was going to be spent in finding errors in my own programs.

Wilkes decided that the laboratory would address the challenge of making programming easier. He assigned the task to David Wheeler. Wheeler had been one of the outstanding mathematicians of his undergraduate years, and he devised a programming system of such brilliance that Wilkes was bowled over. He called it a tour de force of programming.

The heart of the programming system was the **subroutine library**. Most programs, it turned out, tended to use a fairly small set of common functions – such as printing a table of numbers, computing a trigonometrical function or integrating a differential equation. By making these functions available as small pre-written programs (that is, subroutines), the programmer would be saved a lot of coding effort and the consequent potential for making errors.

At that time program debugging was done by sitting at the control desk of the machine and executing the program instruction by instruction, observing the contents of the memory on **monitor tubes**. This was a slow process, which also inconvenienced other users who might be waiting their turn to access the computer. The laboratory came up with two programming aids to avoid the need for debugging at the machine – *post-mortems* and *checking routines* (the latter invented by Stanley Gill). These programs enabled diagnostic information to be printed that could be studied by the programmer at leisure, away from

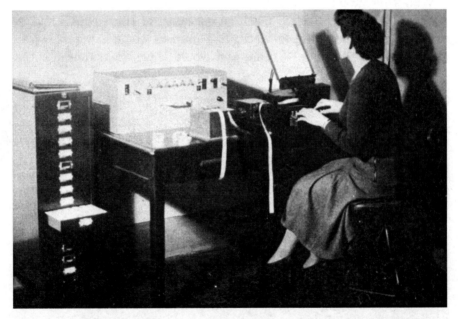

The subroutine library   The heart of the EDSAC programming system was the subroutine library. In order to make life easier for the programmer, common functions (such as printing a result, or calculating a square root or a cosine) were available as pre-written library subroutines. This not only saved the programmer effort in writing programs; it also reduced the potential for making errors.

Library subroutines were kept on punched paper tape and stored in the steel cabinet visible at the left of the photograph. When it was necessary to include a subroutine in a program it was mechanically copied on to the user's main program tape and then returned to the cabinet.

Subroutines are one of the great programming inventions, independently invented by Turing, the Moore School computer group, and others. They are still a cornerstone of software technology.

The EDSAC monitor tubes, which enabled programmers to observe the contents of memory while a program was running

the machine. These techniques were adopted, or reinvented, almost everywhere and became known as *dumps* and *traces* respectively.

With this programming infrastructure, by early 1950 it became possible to have a full-time machine operator for the EDSAC and to provide a true **computing service**. Programmers would bring their program tapes to the EDSAC room and would leave them there for the

The EDSAC computing service in about 1950

Professor Sir John Kendrew, who used the EDSAC to determine the molecular structure of myoglobin, for which he shared the Nobel Prize for chemistry in 1962

operator to process and return for the results later in the day. This greatly improved the utilisation of the machine.

WILKES, WHEELER AND GILL

In September 1950 the programming system was described in a *Report on the Preparation of Programmes for the EDSAC and the Use of the Library of Sub-routines*. Wilkes sent copies of this report to all the computer laboratories he knew about. So far, no other machines had been completed (apart from those at Manchester – see Chapter 4), and it was only natural that most people designing programming systems looked to the Cambridge model.

In the spring of 1951 the report was published as the classic textbook *The Preparation of Programmes for an Electronic Digital Computer*. This was the first textbook on programming and was usually known as 'Wilkes, Wheeler and Gill' or simply 'WWG'. The book was highly influential and was perhaps the single most important outcome of the early years of the Mathematical Laboratory.

THE LAST DAYS OF THE EDSAC

The EDSAC was the beating heart of the Mathematical Laboratory for nearly a decade. From 1953 the laboratory organised summer schools in programming for newcomers to the computer scene. Most of the students went on to play important roles in the early academic and industrial institutions of computing. The EDSAC was used by researchers throughout the university. For example, the molecular biologist **John Kendrew** used EDSAC for the calculations in investigating the molecular structure of the protein myoglobin for which he won a Nobel Prize. This was a stepping stone towards the discovery of DNA.

But EDSAC's days were always numbered. Computer technology was maturing rapidly and EDSAC 2, a new machine 50 times faster than the original, was waiting in the wings. EDSAC was finally closed down on 11 July 1958.

4
THE MANCHESTER MACHINES

Christopher Burton and Simon Lavington

MEMORIES ARE MADE OF THIS ...

Whilst mercury delay-line storage systems were being developed at NPL, Cambridge and elsewhere, another line of research had been started in America. Engineers at the Massachusetts Institute of Technology (MIT) had attempted to memorise, or store, a radar scan using electrostatic charges inside a cathode ray tube (CRT). These experiments were seen by F C (Freddie) Williams, one of the key British electronics designers at the Telecommunications Research Establishment (TRE) at Malvern. Williams was invited to visit the United States in late 1945 and in mid 1946 to help write a series of books describing all the developments in the radar field that had emerged during the Second World War. While there he saw the MIT experiments. He also visited the ENIAC computer at the Moore School and discussed with Presper Eckert the possibility of using the CRT technique to store digital data. The US engineers were discouraging, but Williams returned to TRE determined to experiment on these lines himself.

By the autumn of 1946 a young member of Williams's team, Tom Kilburn, was helping with the CRT experiments at TRE. It is interesting that, though Freddie Williams and Tom Kilburn later claimed they knew nothing about computers, they had picked up enough understanding to visualise the way a computer might function and the role of the storage system within it. A possibility is that John Womersley, head of the Mathematics Division at the National Physical Laboratory (NPL), was their mentor, because he had discussions with Williams at TRE in August 1946 and again at NPL in November of that year.

By the autumn of 1946 **Williams and Kilburn** had built a memory system at TRE that could store a single binary digit. This was a long way from the many thousands of digits needed for a computer, but it was a promising start – especially as their technique seemed to offer *random* (as opposed to sequential) access to information.

Freddie Williams and Tom Kilburn Professor Sir Frederic Williams KB, FRS (1911–77) and Professor Tom Kilburn CBE, FRS (1921–2001) would have described themselves as engineers who knew very little about stored-program computers when they both arrived at the University of Manchester in January 1947.

Williams had gained an engineering degree at the University of Manchester in 1932 and then a scholarship to the University of Oxford, where he obtained a DPhil in 1936 for research on vacuum tube circuits. He returned to lecture at Manchester and was awarded a DSc in 1939. The war years were spent at TRE, where Williams led a central circuit-design team responsible for many of the electronic innovations that contributed to the success of British wartime radar. He returned to Manchester after the war to become head of the Department of Electrical Engineering. There he continued innovating – both in computer hardware and in other areas of engineering such as variable-speed alternating current (AC) motors – right up to his death.

The photograph shows Tom Kilburn (left) and Freddie Williams at the control keyboard of the Manchester University Mark I computer in 1949.

Kilburn graduated in mathematics from the University of Cambridge in 1942 and, following a crash course in electronics, was sent to join Freddie Williams's group at TRE. After the war Kilburn was seconded to Manchester to help perfect William's design for CRT storage, gaining a PhD in 1948. From then on, until his retirement in 1981, Kilburn devoted his life to the design of high-performance computers of which the fifth, called MU5, came into operation in 1972. All five machines had industrial derivatives, of which the fourth (the Ferranti Atlas) was perhaps the most outstanding in its time. In 1964 Kilburn founded the Department of Computer Science, which grew out of the Department of Electrical Engineering, at Manchester.

At this time Williams was appointed to be head of the Department of Electro-technics (now Electrical Engineering) at the University of Manchester, where he moved at the beginning of 1947. It is likely that the selection committee effectively 'head-hunted' him, because a member of the committee, Max Newman, was anxious to have an electronic computer built and was aware of Williams's work at the leading edge of storage system design. Newman had left Bletchley Park and become Professor of Mathematics at Manchester in October 1945. His experience of wartime code-breaking and his close contact with Turing and his ideas about computation made him determined to set up a project to purchase or build a high-speed general-purpose computer to investigate mathematical problems. This was, as he explained in a letter to von Neumann in February 1946,

> before I knew anything of the American work [EDVAC], or of the scheme for a unit [the ACE] at the National Physical Laboratory.

Newman himself had no relevant engineering ability but assumed that he would be able to buy a computer as soon as one with a viable storage system became available. In July 1946 he had obtained a large grant from the Royal Society to set up a Computing Machine Laboratory at Manchester and had brought two mathematics colleagues, David Rees and Jack Good, from Bletchley Park to staff it.

But this was a notional laboratory – no actual room or computer yet existed in 1947. Getting Williams to come to Manchester with his storage research and creative engineering skill appeared to offer an ideal way of getting the central problem of a computer, the memory, sorted out. Williams's former employers at TRE were happy for this research to be continued at Manchester. They provided materials and components and, importantly, seconded Tom Kilburn to the university to maintain the momentum of the project.

The two engineers arrived at Manchester University, where Williams, as head of the Department of Electro-technics, soon became involved with administration. Meanwhile, Kilburn brought the storage equipment from Malvern and set it up in a spare laboratory, the former 'Magnetism Room'. He made rapid progress in refining the technique for **storing digits as charge patterns on the face of the CRT**. In the summer of 1947 another colleague from TRE, Geoff Tootill, was seconded to help with the storage project. By the autumn of 1947 they were able to store 2,048 binary digits for a matter of hours. The equipment consisted of several 2.2 m tall racks, each holding a dozen or so shelves containing the electronics.

At this point (December 1947) Kilburn wrote an important report describing the Williams–Kilburn storage system. The report was widely circulated in the UK and the USA – and a copy even appeared in Moscow! Throughout this development phase the two engineers had been careful to protect their invention with patents – indeed, IBM in the United States subsequently took out **licences to use the system in the IBM 701 and 702 computers**. At the end of 1947, however, Kilburn was preoccupied with the one vital question: 'How well will the store work in the hurly-burly of high-speed computing?'

To answer this question, Kilburn decided that the simplest thing to do was to build a minimal computer incorporating the CRT storage system, using the equipment and resources still being provided by TRE. He was by now familiar with the principles of a universal computer – indeed, the basic concepts were not hard to understand. Furthermore, at some stage in 1947 Newman's mathematicians had explained their specification of a machine they would like for their Computing Machine Laboratory. However, in 1947 Williams and Kilburn still had to prove the effectiveness of their novel storage system and were not yet ready to embark on developing any large-scale computer. A small engineering prototype was first required.

CRT storage In a cathode ray tube (CRT) a focused beam of electrons causes a glowing spot on the inner phosphor coating of a screen at the far end of the tube. At the same time the area of the spot acquires an electrostatic charge from the electron beam. The charge gradually leaks away but can be regenerated by external circuits. The presence or absence of a charged spot at any particular place on the screen can also be detected by external circuits, via capacitive coupling to a metal plate on the outside of the tube face. A pattern of many spots can be 'written' onto the screen and can be detected or 'read' by the external circuits.

Each spot, or the absence of a spot, can be made to represent a binary digit. Other schemes, such as a focused and a defocused spot or a dot and a dash, may also be used to indicate binary zero and one. By deflecting the CRT's electron beam, each spot in an array of spots can be quickly scanned or 'addressed'.

Since the time to address any spot is independent of its physical position in the array of spots, a CRT storage system is said to be a 'random-access memory' device, similar to a modern RAM. The photograph shows a program stored as dots and dashes on the screen of the Store CRT of the replica SSEM ('Baby' computer). In each line, the least-significant digit is on the left. The top line is bright because it is being accessed 32 times more often than the other lines.

A cabinet of Williams–Kilburn CRT storage tubes, as used under licence by IBM in the IBM 701 and 702 computers from about 1953 onwards, with F C Williams (on the left), H J Crawley (National Research Development Corporation) and J C McPherson (IBM)

THE BABY COMPUTER

Kilburn and Tootill got down to designing the additional circuits needed to make the storage system into a small computer, to be called the Small-Scale Experimental Machine (SSEM) or 'Baby'. To save effort, Kilburn decided to provide only subtraction in the arithmetic unit, on the grounds that addition could be achieved by double negation, whereas an adder cannot subtract. He included a total of only seven different orders in the instruction set, a subset of those suggested by Newman's mathematicians but sufficient for universality. The store was reduced to 1,024 digits, arranged as 32 words of 32 bits each, to guarantee adequate reliability.

By June 1948 the **SSEM or 'Baby' computer** was complete. Two **simple test programs** were written, one to find the highest common divisor of two numbers and the other to find the highest factor of a number. The data was chosen so as to give long run times of about 50 minutes, involving the execution of about 3.5 million instructions. To test the machine, a program was inserted into the store via push buttons and switches. For some days each attempt at a run resulted in failure, usually due to a wiring error. Then, on the morning of Monday 21 June 1948, the highest factor program gave the correct result. They tried another run, and again it was correct. What excitement! 'Quick!' said Tom Kilburn to Geoff Tootill, 'Go and fetch Freddie!' Williams was fetched from his office, and for a third time the program ran correctly. That was a momentous day: the first time ever in the world that a universal stored-program electronic computer had successfully worked!

The Baby was indeed a simple machine. Only seven operation codes (functions) were available, as explained in Appendix A. The random-access store had a capacity (in modern terms) of only 128 bytes. Nevertheless, in July 1948 Sir Ben Lockspeiser, Scientific Adviser to the Ministry of Supply, visited Manchester to see the Baby. He was greatly impressed. He quickly arranged for government funding to be made available for the Manchester engineering company Ferranti Ltd to make a properly engineered version of the university machine, the only contractual specification being that it should be made '... to the instructions of Professor F C Williams'.

Today a full-size working replica of the Baby may be seen at the Museum of Science and Industry in Manchester.

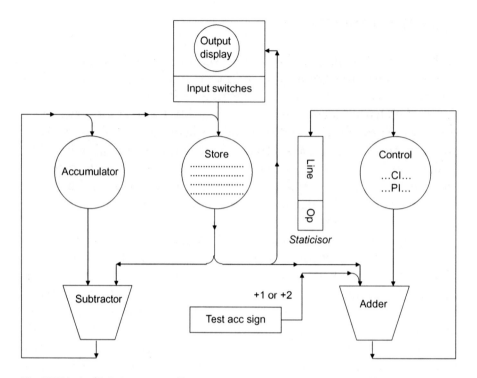

The SSEM: the 'Baby' computer This diagram shows a simplified schematic of the SSEM.
The machine contained three Williams–Kilburn CRTs. One was used for the main memory, S, of 32 words each consisting of 32 bits. A second CRT held the 32-bit accumulator, A. The third was the 'control' tube, C, which held the Present Instruction, PI, and the Program Counter, CI. (On a modern computer the accumulator, PI and CI are all known as *registers*, or *short-term* stores. In particular, the accumulator is a special register whose main task is to hold the number resulting from each individual instruction immediately after it has been obeyed.)

The box labelled 'Staticisor' in the diagram was a set of eight flip-flops: three for the operation bits and five for the operand address, which together made up an instruction. The full list of SSEM instructions is explained in Appendix A.
A program and its data had to be inserted into the SSEM's store manually, via the input switches. The results of the SSEM's calculations could be read as dots and dashes appearing on the output display. A fully functional, faithful reconstruction of the Baby can be seen today at the Museum of Science and Industry in Manchester.

THE BABY GROWS UP

Over the next year the Baby was successively enhanced and was renamed the Manchester University Mark I – sometimes also known as MADM (Manchester Automatic Digital Machine). The team's main aim was to try out new ideas and new hardware, so as to provide Ferranti with a realistic prototype on which to base their commercial version. A related objective was to give Professor Newman's mathematicians experience in using an authentic machine for their mathematical problems.

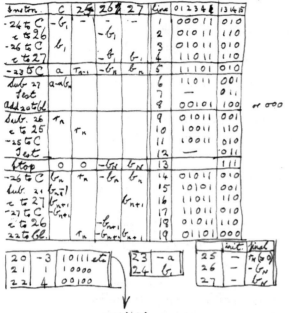

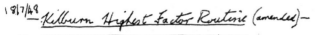

The amended first program to run on the 'Baby', on the morning of Monday 21 June 1948 – from Geoff Tootill's laboratory notebook

The Manchester University Mark I computer with its design team in June 1949 (left to right): Dai Edwards, Freddie Williams, Tom Kilburn, Alec Robinson and Tommy Thomas; (inset) Geoff Tootill

Besides Williams, Kilburn and Tootill, the **Manchester University computer design team** was expanded in the autumn of 1948 by taking on three postgraduate engineering students: Alec Robinson, Dai Edwards and Tommy Thomas. Alan Turing, who had effectively left NPL in 1947, joined Max Newman's mathematicians in October 1948 with the job title Deputy Director of the Royal Society Computing Machine Laboratory. His salary was the first call upon Newman's Royal Society grant.

It was Turing who took the lead in developing programming systems for the Manchester University Mark I. Having been familiar at Bletchley Park with the five-bit international teleprinter code, he arranged for standard five-track teleprinter paper tape equipment to be connected to the Mark I computer for input and output. More information on Mark I programming is given in Appendix A.

The computer's word length was increased to 40 bits, thus making it a convenient multiple of five bits. Each word represented either a (fixed-point) number or two 20-bit instructions. The instruction set was expanded from seven operations to 26, including hardware multiply. Most significantly, there were two index, or modifier, registers, called 'B lines' – for which the patent was dated 22 June 1949. Modifier registers are seen on every modern computer. An example of their use is given in Appendix A.

The mathematicians wanted a very large-capacity store, beyond what was economically reasonable using CRTs. The team turned to the idea of using a magnetic drum, as suggested by Andrew Booth at Birkbeck College (see Chapter 6). The **Manchester drum** stored up to 2,048 40-bit words. The primary store (RAM) consisted of two Williams–Kilburn tubes, each holding 64 40-bit words. Data could be transferred between the fast Williams–Kilburn store and the drum, at first manually and then, from October 1949, under program control.

By the autumn of 1949 the Manchester University Mark I had become a relatively powerful facility. Being a prototype that had 'grown like Topsy', it did not have great reliability – though one unusually long error-free run of 17 hours was recorded in June 1950. Max Newman's group suggested that the computer should be used to investigate Mersenne prime numbers and wrote a suitable program. Other problems tackled in the period 1949–50 included an investigation of the Riemann hypothesis (the Zeta function) and calculations in optics (ray tracing). In October 1949 Audrey Bates was taken on as the first mathematics research student to use the computer. Her supervisor was Alan Turing, and the title of her Master's thesis was *On the mechanical solution of a problem in Church's lambda calculus*. She did not continue

The prototype Manchester drum store in 1949. Because of its squat shape, the nickel-plated drum was referred to locally as the 'magnetic wheel'. It was situated in a laboratory above the main computer room, giving rise to the terminology of transferring pages 'down from', or 'up to', the drum.

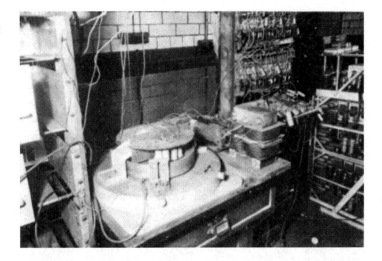

to a doctorate, one reason being a remark by Professor Newman in December 1949, who said: 'Miss Bates, I don't think anyone will ever get a PhD involving computers.'

The Manchester Mark I machine was finally dismantled in the autumn of 1950. The mathematical users were now anticipating the delivery of the fully engineered production version, to be called the Ferranti Mark I. The university engineers, led by Tom Kilburn, turned their attention to the design of a new research machine that would be ten times faster. Manchester was by now firmly established as one of the key centres in the world for computer engineering.

FERRANTI ENTERS THE PICTURE

The manufacture of a fully engineered version of Manchester University's prototype computer began in earnest at Ferranti Ltd at the end of 1949. The design of the **Ferranti Mark I** naturally incorporated all that had been learned during the operation of the university prototype. In a report for Ferranti dated November 1949, Geoff Tootill described 'the requirements placed upon us by Professor Newman and Mr Turing'. One special instruction they called for was a 'population count', or 'sideways add'.

> The sideways addition facility is a requirement of Professor Newman's who, it is believed, proposes to use it for investigating various propositions in symbolic logic. The truth or falsehood of these propositions will be denoted by the binary digits 1 or 0 and the truth or falsehood of 20 propositions will be stored on one line. It is then required to know how many of these propositions are true ...

The Ferranti Mark I computer with the operator's control console between the rows of cabinets holding the logic circuitry. The person standing to the right of the console is Alan Turing.

Another new feature was a special hardware random-number generator, suggested by Alan Turing. Christopher Strachey (see Chapter 5) made a light-hearted use of the Ferranti Mark I's random-number generator in the summer of 1952 for 'creating' love letters, of which the following is an example.

> Darling Sweetheart,
> You are my avid fellow-feeling. My affection curiously clings to your passionate wish. My liking yearns to your heart. You are my wistful sympathy; my tender liking.
> Yours beautifully,
> M.U.C.

The Ferranti Mark I arrived at the university on 12 February 1951, thereby becoming the first production computer to be delivered anywhere in the world. In March 1951 Alan Turing wrote the first *Programmers' Handbook* for what was initially, and confusingly, called the Manchester Electronic Computer Mark II. (It was not until about a year later, once the marketing possibilities had become clear, that the name was changed to the Ferranti Mark I). To celebrate the arrival

The magnetic drum store for the Ferranti Mark I computer

of the new computer, a special Inaugural Conference was held at the university from 9 to 12 July, attended by 169 people from a total of 12 countries. Alan Turing presented a paper entitled 'Local programming methods and conventions'. This, as it turned out, was Turing's final public appearance at any computer conference.

A SUPERCOMPUTER

The Ferranti Mark I machine was, in 1951, probably the most powerful scientific supercomputer available. Yet, 60 years later, we might easily dismiss it as a weakling. In modern terminology it had a primary memory (RAM) of only about 1 Kbyte and a secondary memory (the **magnetic drum**) of up to 16 Kbytes. More precisely, the Ferranti Mark I's primary store consisted of eight Williams–Kilburn CRTs, each holding 64 half-words of 20 bits each. Each block of 64 half-words was called a page. The drum had a maximum of 256 tracks, each track storing two pages. Each page, when stored on the drum, also had a 65th line containing its drum address. The 65th line was the germ of an idea that developed into the modern concepts of paging and virtual memory, when implemented on the joint Ferranti and Manchester University computer called Atlas in 1962.

The main central processor and storage units of the Ferranti Mark I were contained in two bays, each 17 ft long by 9 ft high (5 m × 2.7 m). They included 1,600 pentode vacuum tubes and 2,000 thermionic diodes and consumed 25 kW of power. The **multiplier section**, which formed part of one of these bays, is shown opposite. Input–output equipment was quite basic, being limited to a paper tape reader (200 characters per second) and a paper tape punch plus teleprinter (printing at 10 characters per second).

PROGRAMS AND USERS

Alan Turing devised a programming system for the Ferranti Mark I that was not for the faint hearted. There was no symbolic assembler (see the sample program in Appendix A). There was an added complication: the hardware engineers displayed bit patterns with time going from left to right so that, when written down, the least-significant digit appeared on the left-hand end of a string of bits. Turing had chosen to perpetuate this convention, both at Manchester and, previously, at NPL, so that programmers were also obliged to write out their bit patterns in 'backwards binary'.

Things got better with the arrival in October 1951 of Tony Brooker, who took over from Alan Turing as chief systems programmer. Brooker had spent two years in the Computer Laboratory at Cambridge, working on software development for EDSAC. At Manchester he devised a scheme that allowed users to write programs in the style of algebraic expressions, with simple ways of implicitly calling standard library subroutines. Brooker's scheme, **Mark I Autocode**, was released in March 1954 and was probably the world's first publicly available *high-level language*. It was about two years ahead of the first Fortran compiler. Manchester programmers had stolen a march on Cambridge!

The multiplier section of the Ferranti Mark I – part of one side of the computer, shown with the cabinet doors removed

Looking back, it seems that the mathematicians who had originally requested facilities such as the 'sideways add' made little use of the Ferranti Mark I computer for investigating 'various propositions in symbolic logic'. In February 1950 Alan Turing started working on his mathematical theory of embryology (**morphogenesis**), for which he used partial differential equations. He became interested in the growth and form of living things and was fascinated by symmetry in nature and by the colour patterns on animals – for example stripes, spots and dappling. In contrast, most other users of the Ferranti Mark I were scientists and engineers working on down-to-earth problems such as the optimum design of turbine blades. Industrial users were charged £20 per hour of computer time, equivalent in 2012 prices to about £400 per hour.

Alan Turing's day-to-day connections with computing staff at the university became much less frequent after 1951, as he immersed himself in the difficult theoretical problems of morphogenesis. His seminal paper, 'The chemical basis of morphogenesis', was published in August 1952. Turing himself rated its importance as equal to that of his 1937 paper 'On Computable Numbers'. For the next 18 months he continued to tackle the harder areas of morphogenesis. Alas, he was destined to leave this work unfinished. On 7 June 1954 Alan Turing was found dead in his house at Wilmslow, south of Manchester, having apparently eaten an apple dipped in cyanide. His death was completely unexpected, coming as a great shock to all who knew him. It is now thought that personal problems, including the difficulties he encountered in the restrictive society of the 1950s concerning his sexual orientation, contributed to the sad end of this genius.

WHAT CAME NEXT?

After delivering a second Ferranti Mark I computer to the University of Toronto in 1952, the Ferranti company made certain improvements to the basic design in the light of user experience. With the financial backing of the National Research Development Corporation Ferranti produced the Mark I* (pronounced 'mark one star'). This computer differed from its predecessor in several minor respects and in two major ones: it had a simplified instruction set, consisting of 30 functions (things such as sideways add and random-number generation were excluded as being little used), and it had extra CRT storage. Seven Ferranti Mark I* computers were delivered to customers between 1953 and 1957.

Meanwhile Tom Kilburn and his computer design team had not been idle. Two projects were being pursued. First a successor to the Mark I, called 'Meg', was built. This was faster and more reliable, consumed less power and, notably, featured hardware floating-point instructions. Meg first ran a program in May 1954. Secondly, a small experimental transistor computer was built at Manchester in order to gain experience of semiconductor technology. This machine first ran a program in November 1953. As far as is known, Alan Turing took no interest whatsoever in these post-1951 computer design activities at Manchester University. He probably

Mark I Autocode The Autocode system used the symbols v1, v2, v3, ... to stand for floating-point variables and n1, n2, n3, ... to stand for integer variables. An array of 100 floating-point numbers could be represented by vn1, where n1 took on the values 1 to 100. Floating-point calculations were automatically performed at run time by interpretive library routines. Other symbols were used for standard library routines for printing, calculating a square root, etc. The symbol j was used to indicate a control transfer (i.e., jumps or branches in a program), and the lines of a program could be labelled. Thus, the Autocode statement

J2, 100 ≥ n1

meant: 'Jump to program line 2 if the user's variable n1 is less than 100.'

By way of example, here is a simple Mark I Autocode program that calculates the root mean square of one hundred real (i.e., floating-point) variables v1, v2, v3, etc.

```
 n1 = 1
 v101 = 0
2v102 = vn1 x vn1
 v101 = v101 + v102
 n1 = n1 + 1
 j2, 100 ≥ n1
 v101 = v101/100.0
 *v101 = F1(v101)
```

The symbol * caused printing to ten decimal places on a new line; F1 signified the intrinsic function 'square root'.

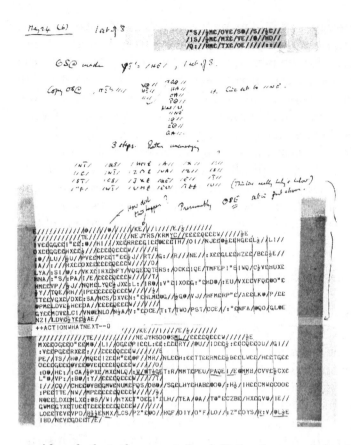

considered that they contained little that was fundamentally new. Although some might agree, most people are thankful that today's computers are more cost-effective and easier to use than those of the 1950s.

5
MEANWHILE, IN DEEPEST HERTFORDSHIRE

Simon Lavington

THE ADMIRALTY'S SECRET

Whilst computer design activity was getting under way at the National Physical Laboratory and at the Universities of Cambridge and Manchester, something rather secret was starting to emerge from a redundant wartime factory at Borehamwood in Hertfordshire. The Admiralty, who owned the factory, had been worried by the modest performance of the Royal Navy's electromechanical anti-aircraft gunnery-control equipment during the war. Accordingly, in 1946 the Admiralty decided to fund the design of a new high-performance electronic *digital* computer to do the gunnery-control job. Since it was to be connected to an advanced real-time target-tracking radar, the digital computer had to be very fast.

With the blessing of the Admiralty the wartime head of naval radar research, John Coales, left government employment to lead the digital project. He recruited a team of scientists and engineers to start work at Borehamwood in the autumn of 1946. His team at the Borehamwood Laboratory was, for organisational reasons, placed under the control of a long-established scientific instrument company called Elliott Brothers (London) Ltd.

By the summer of 1950 Elliott's Borehamwood Laboratories had managed to get the first of their **secret computers**, a high-speed digital gunnery-control computer called the Elliott 152, to work under test conditions. Soon afterwards, however, the Admiralty cancelled the contract – one reason being that, by 1950, surface-to-air guided weapons seemed to offer a more promising defence of ships against aircraft. Though Elliott's secret gunnery-control computer never saw service at sea, the reputation that John Coales's group had built up for state-of-the-art digital electronics was not wasted.

Secret Elliott computers The Elliott company's research lab at Borehamwood built four classified general-purpose computers between 1950 and 1956. A fifth special-purpose code-cracking computer called OEDIPUS, built for GCHQ in 1954, was so secret that there is no surviving photograph of it in the public domain.

The Elliott 152 prototype computer, intended for real-time online naval gunnery control (1950).

The Elliott Nicholas computer, used for guided bomb trajectory calculations for the Royal Aircraft Establishment, Farnborough (1952).

The Elliott 403 computer, called WREDAC, which was delivered to the Long Range Weapons Establishment in South Australia for the analysis of Woomera's missile test-firing data (1956).

The Elliott 153, for rapid plotting of Direction Finding (DF) signal intercepts at GCHQ's Irton Moor Establishment near Scarborough (1954).

Soon three more new defence contracts had come to Borehamwood, each of them resulting in the design and construction of a general-purpose digital computer for classified tasks. One of the secret tasks was related to GCHQ's intelligence activities. The other two were for the design and analysis of guided weapons. These computers were known respectively as the Elliott 153, 403 and Nicholas. A fourth computer called OEDIPUS was also built, though this was a special-purpose digital cryptanalysis machine with – interestingly – large amounts of semiconductor associative (i.e., content-addressable) storage. These four new computers came into operation between 1952 and 1956.

What connection, if any, did all this secret Borehamwood activity have with the other early computer design teams in Britain and America? It is difficult to come up with a precise answer because very few Elliott company records have survived. Norman Hill, who had worked on government research during the war, was a mathematician who led the small Theory Group at Borehamwood. In retirement many years later he remembers the events of 1946–9 as follows.

> It was evident that the computing techniques [required for real-time gunnery control] needed to be extremely fast and accurate, and digital methods were therefore proposed …
>
> Much interest was generated in ENIAC worldwide, and Mr Coales directed us to study the techniques embodied in it. Many visits were made to view ENIAC [by other British researchers], notably by Maurice Wilkes from Cambridge, Wilkinson from NPL, Prof. Hartree from Manchester [later at Cambridge University] and Uttley from TRE. These and other people decided to build digital computers at their various establishments. Our problem [at Borehamwood] was how to seek information on these various plans and designs bearing in mind that we could not reveal our purpose since we were working on a secret contract. I remember going with John Coales in his vintage car to the NPL for discussions with some of the above-mentioned people together with the great Turing who insisted on running to the canteen for lunch in pouring rain …
>
> On another occasion we went to Cambridge. Maurice Wilkes at Cambridge held a series of colloquia in the Mathematical Laboratory every two weeks at which experts on digital computing techniques were invited to give a talk followed by free discussion and tea and buns. As a forum for exchanging information these colloquia were invaluable and they greatly contributed to the growth of knowledge in this new and exciting field. We were able to keep up to date with progress on the different machines which were

49

being designed and built. At about this time [Autumn 1948] Bill
Elliott arrived at Borehamwood to lead the Computing Division. He
had numerous friends at Cambridge, TRE and other places.
By attendance at Cambridge colloquia and visits to other
establishments including those in the USA, we somehow kept in
touch with worldwide developments in computing techniques.

It is certain that the ENIAC interest would have led Borehamwood
to the *EDVAC Report* and to John von Neumann's ideas, especially in
view of the close links between John Coales and Cambridge University.
Maurice Wilkes, who became an Elliott Consultant, provided an ongoing inspiration for Borehamwood. Bill Elliott, Head of the Computing
Division at Borehamwood from 1948 to 1953, had been a good friend
of Maurice Wilkes ever since they had shared an interest in amateur
radio as undergraduates.

INNOVATIONS AT BOREHAMWOOD

Whatever ideas Borehamwood may or may not have got from America
and Cambridge, the designers of the Elliott computers did not choose
the storage technology used by EDSAC. Instead of using mercury
delay lines, Borehamwood at first adapted the Manchester invention of
CRT (electrostatic) storage because it seemed to offer higher speeds –
see Chapter 4. Then in 1952 one of several **Elliott technological
innovations** was introduced, when the team invented its own delay-line
storage technology based on the magneto-strictive properties of nickel.
The magneto-strictive effect enabled electronic pulses to be sent as
acoustic shock waves along nickel wires, just as pulses could be sent
as acoustic waves along tubes filled with mercury. Nickel delay lines
were more robust and cheaper per bit than the mercury delay-line technology. The nickel delay lines were also more robust and more reliable
than CRT storage, but more expensive per bit.

There were other ways in which the early Borehamwood computers did not follow the designs of other pioneering groups. Since real-
time gunnery control demanded high computing speeds, the engineers
were obliged to devise a low-level architecture that permitted all of the
functional units inside the computer, for example the addition circuits,
the multiplier circuits and the temporary storage registers, to operate
in parallel at the same time. From the programmer's viewpoint this
meant that each machine instruction was quite complicated, but it also
meant that each instruction could do the job of perhaps three or four
of the instructions seen in other computers of the period. In their two
earliest secret computer designs Borehamwood computers also had

Elliott technological innovations The needs of the Elliott secret computers led to two interesting general technologies. The first was a set of standard logic packaged circuits, each the size of a paperback book, which were the building blocks from which a family of computers could be built. The second Elliott innovation was the development of the nickel delay-line store in 1952 as an alternative to other early computer memory technologies.

The photograph shows standard packaged circuits, as used for the Elliott 401, 402, 403 and 405 series of computers in the 1950s.

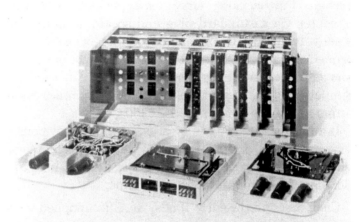

The device at the top left in this photograph is a CRT Williams–Kilburn storage tube from a Ferranti Mark I computer. In the centre is a mercury delay line from an English Electric DEUCE computer, where acoustic signals are transmitted down one tube and reflected back along the other tube. At the right is a package from a Ferranti Pegasus computer containing a nickel delay-line register.

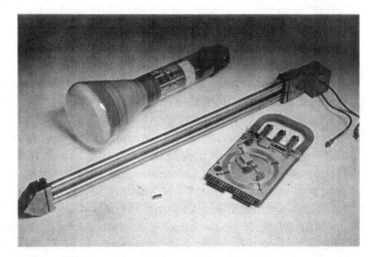

separate stores for instructions and for data. All this is not to say that Borehamwood was unique, but only that the engineers there were very innovative. As a result of this, the multiplication time of 60 microseconds achieved by the Elliott 152 gunnery-control computer was, at the time (1950), the fastest in the world.

Another innovation, and one of longer-lasting significance, was the Borehamwood practice of building a computer's central processor from many individual modules or packages, each package chosen from a limited number of types. Central to this philosophy was the idea of a family of standard logic circuits, robustly packaged, from which a designer could assemble the many functional sub-units that went to form the complete computer. These packages, or building blocks, were carefully specified so that their properties were capable of mass production. Furthermore, since each package was connected to the computer via a standard plug and socket, a faulty package could be readily replaced. At Borehamwood the packages took the form of printed circuit boards with components mounted on one side, each complete package being about the size of a normal paperback book. Similar-sized packages were soon adopted by other computer manufacturers in the late 1950s and the 1960s.

If all of the foregoing innovations seem to leave out any mention of software, this is a reflection of the way most computer manufacturers operated in the early 1950s. To quote Dina Vaughan, one of the few Borehamwood programmers of the time:

> Computers [in the early 1950s] were first applied to mathematical work, and the user did his own programming entirely. Later a small library of common mathematical functions was organised or supplied by the manufacturer. The users were scientifically minded people, usually with considerable mathematical training. When computers started to be applied to business data processing and other non-mathematical fields, the users were seldom scientifically inclined and hence expected much more support from the computer manufacturers ...
>
> The manufacturers had to devote more and more manpower and cost to application programming, all of which had to be recovered in the price of the equipment as the customer expected the programming support 'free'. From the customer end, not only was there an increasing shortage of trained people, but it became doubtful whether teams below a certain size were really viable.
>
> This is an environment where experience is at a premium and must be spread, but job mobility is high with more loyalty given to the programming world at large than any particular user's business.

Dina Vaughan, who later married Elliott's Managing Director of Computing, Andrew St Johnston, left Borehamwood in 1958 and, in February 1959, founded the UK's first software house to address the above problem.

Bill Elliott and Andrew St Johnston The task of overseeing the transfer of Borehamwood's computer technology from defence-related projects to the open market was headed at first by W S (Bill) Elliott – who was no relation to the founder of Elliott Brothers (London) Ltd – and then by Andrew St Johnston.

Bill Elliott (1917–2000) graduated in Physics from Cambridge in 1938. His postgraduate research was interrupted by the war, during which he worked on army radar at the Air Defence Research and Development Establishment. He joined the Borehamwood Laboratories to set up a Computing Division towards the end of 1948, and left in 1953 to join Ferranti Ltd, where he led the team developing the Pegasus computer. He moved to IBM in 1956 to head IBM's new UK Laboratories at Hursley. He moved in 1961 to Cambridge University to manage a large computer project and from there in 1966 to Imperial College, London, where he became Professor of Computing. He retired in 1982.

Andrew St Johnston (1922–2005) graduated in Electrical Engineering from Imperial College in 1943 and became a radar officer in the Royal Navy during the war. St Johnston joined the Computing Division at Borehamwood in 1949 to work on the Elliott 152 computer. From 1953 to 1968 he was the General Manager of the Computing Division, responsible for the Elliott 400 series of computers and all those that followed (the Elliott 500, 800, 900 and 4100 series) until the takeover of Elliott-Automation Ltd in 1968. From 1968 to his retirement in 1999 he managed the software company Vaughan Systems Ltd.

SWORDS INTO PLOUGHSHARES

Pulling all of Borehamwood's defence-related technology through to the civil marketplace became an ambition for the National Research Development Corporation (NRDC). It was NRDC who provided the encouragement and finance for the team, headed successively by **Bill Elliott and Andrew St Johnston**, to – so to speak – turn Borehamwood's swords into ploughshares. This is how it happened.

Towards the end of 1950 NRDC was predicting a need for a modest computer that would complement the high performance – and high price – of the forthcoming Ferranti Mark I. The Ferranti Mark I, described in Chapter 4, was first delivered in the spring of 1951. Its open-market price was about £95,000 (equivalent to perhaps £2 million at 2012 prices). NRDC foresaw another market opportunity for an easy-to-install, easy-to-maintain computer that would sell at about a quarter of the price of the high-performance Ferranti machine. Borehamwood's family of robustly packaged digital circuits appealed to NRDC, who approached the Elliott management with a view to developing a new and relatively cheap computer.

The result was a machine called the **Elliott/NRDC 401**. This was demonstrated to the public at the Physical Society Exhibition, held in London in April 1953. The 401 is believed to be the first computer to have been put on public display at an exhibition site. It aroused much interest – especially since few people outside the secret world of defence contracts were aware that the apparently sedate company of instrument makers, Elliott Brothers (London) Ltd, based in Lewisham (south London), had a laboratory 20 miles away in rural Hertfordshire that specialised in digital electronics.

NRDC was responsible for providing more than just money for the 401 development. **Christopher Strachey**, a Harrow schoolmaster and enthusiastic amateur programmer, had learned to program both the Pilot ACE computer at the National Physical Laboratory and the Ferranti Mark I computer at Manchester. Such was his skill at programming the Manchester machine that

The Elliott 401, which was demonstrated at the Physical Society's annual exhibition in London in April 1953 – probably the first time a computer was put on public display at a trade show. This machine marked the start of the company's production of open-market computers.

Alan Turing recommended NRDC to employ him as a consultant. Thus it was that, from November 1951, Strachey was to offer advice to Borehamwood on the instruction set for the 401 computer. He was one of those rare individuals who had a feel for both the theoretical and the practical sides of computer hardware and software. He went on to take a prominent role in the design of the Ferranti Pegasus computer, which, by 1956, was to become an outgrowth of the ideas first demonstrated in the Elliott 401.

The birth of the 401 computer was not achieved without a degree of pain and anguish within Borehamwood. Elliott Brothers (London) Ltd was financially weak at the time, prompting the management to seek to reduce staffing levels. In May 1953 Bill Elliott left to start a rival computing project at the London laboratory of Ferranti Ltd, the project becoming known as Pegasus. On his departure leadership of the Borehamwood Computing Division passed to Andrew St Johnston, an ex-naval radar officer who had joined the company in 1949 to work on the Elliott gunnery-control computer. St Johnston was to direct all of Borehamwood's many computing projects until he left the company in 1968, the year in which ICL was founded (see Chapter 7).

Following the 401, Borehamwood developed an improved production version called the Elliott 402.

Christopher Strachey There were few computer pioneers who seemed to be equally at home with computer hardware, software and theory. Christopher Strachey (1916–75) was one. A skilled programmer, he also made contributions to the hardware design of the Elliott 401 and Ferranti Pegasus computers whilst working for the National Research Development Corporation (NRDC). He went on to become Oxford University's first Professor of Computer Science, specialising in denotational semantics with the aim of proving programs correct. In some ways Strachey's later work was a practical outgrowth of Alan Turing's 1936–7 theoretical paper 'On Computable Numbers'.

Ten of these machines were delivered to customers in the period 1955–9. Two of these were exported (to France and to Germany). Most of the end-user applications were in science and engineering. Then came the Elliott 405 computer. This was a much larger machine that was specifically aimed at the commercial data-processing arena. Interestingly, the structure of the 405 was strongly influenced by Borehamwood's experience in designing a secret computer, known as the 403, for the Long Range Weapons Establishment at Woomera in Australia. The Elliott 403 had to handle the input and output of large amounts of data arising from the many Woomera flight-trials of guided missiles. This required the computer to be equipped with magnetic tape decks, a fast line printer and four graph plotters. To handle the high data rates, the 403 had a special input–output processor that was almost as large as the main computer. The Elliott 405 handled even larger amounts of data for business applications.

Thirty-three of the Elliott 405 machines were delivered between 1956 and 1962. The first customer was **Norwich City Council** – probably the first local authority to purchase a computer. Between 1956 and 1967 Elliott had an agreement with the National Cash Register Co. Ltd (NCR), whereby NCR became responsible for marketing the Elliott 405 and other Elliott products to the commercial data-processing sector. Although NCR's track record in electromechanical accounting machines was thought by some to be an unsuitable background for promoting the latest electronic stored-program computers, Borehamwood's relatively small Computing Division was not able to undertake commercial marketing without help, and NCR's experience of day-to-day business practices was thought to be just what was needed.

THE COMING OF AUTOMATION

Nowadays a production run of only 33 computers seems ridiculously small. However, by the end of 1955 a total of fewer than 20 production computers had been designed and delivered by all the British computer manufacturing companies put together. Five of these machines had been exported and the remainder installed in the UK. Not long afterwards American-built computers – notably those of IBM – began to appear in Britain, and things changed for ever. A review of the early British companies, their products and the end-user applications is given in Chapter 7.

For Elliott, the mid 1950s was the beginning of an exciting time. The company's Managing Director, **Leon Bagrit**, believed passionately

Norwich City Council's Elliott 405 computer at its inauguration in 1956. This is believed to have been the first computer purchased by local government. Dina Vaughan, who wrote the initial applications software, is seen on the right of the photograph.

Leon Bagrit arrived from the Ukraine as a 12-year-old refugee in 1914, not speaking a word of English. By 1947 he had become the Managing Director of the long-established scientific instrument company Elliott Brothers (London) Ltd. He successfully transformed this ailing firm into one of the UK's leading suppliers of computer equipment, renaming it Elliott-Automation. Bagrit was knighted in 1962 and gave the BBC Reith Lectures in 1964, by which time he was known in the media as 'Mr Automation'. He retired from being chairman of GEC-Elliott-Automation in 1973 and died in 1979.

in industrial process control. With Elliott's newly acquired ability to design and sell computers to both scientific and business organisations, Bagrit realised that digital computers could also revolutionise industrial process control. What had previously been a modest data-recording and instrumentation exercise was to be transformed into what was soon to be called *automation*. Bagrit's vision was that digital computers would be widely used for looking after all sorts of manufacturing processes, making decisions and adjustments that had previously been carried out by human operators. The company name was changed to Elliott-Automation in 1957. Elliott made 50 per cent of all the new computers sold in the UK in 1961. Many of these were destined for industrial control.

6
ONE MAN IN A BARN

Roger Johnson

X-RAY CALCULATIONS

The story of the development of the UK's best-selling early computer used for commercial IT starts in the unlikely world of the crystal structure of explosives.

In the closing days of the Second World War a prominent British scientist, J D Bernal, was planning his return from war service to the quieter world of academia. He held the Chair of Physics at Birkbeck College, London University, and was planning to form a group of academics to examine the structure of crystals using X-rays, work which contributed in the 1950s to the discovery of the double helix. This involved solving large sets of equations, which, before computers, had to be done largely by hand using mostly simple electromechanical calculators. It took weeks to complete one set of calculations.

Bernal's attention was drawn to a young academic called **Andrew Booth**, who had worked on the X-ray structure of explosives during the war. In 1975, in an interview for the Science Museum in London, he related how during the war he managed a small team of women doing these calculations, and recalled that:

> being by temperament a mathematician I don't like arithmetic ... I didn't think much of the methods they were using and I tried to do two things. In the first place, I devised some better mathematical methods ... but I also made one or two small hand calculators.

Andrew Booth proved an excellent choice: he held a PhD in crystallography and was a talented mathematician, but, most importantly for computing, he was described by those who worked with him as a 'natural engineer'. (His engineering skills were perhaps acquired from his father, who was a marine engineer and part-time inventor.) Consequently it was not surprising that he started building analogue and other mechanical devices to reduce the need for laborious calculation by hand. However, shortly after his arrival at Birkbeck he also started to build his first electromechanical digital computer, the **Automatic Relay Calculator (ARC)**. This was a special-purpose computer to solve sets of crystallographic equations. The design involved using 600 relays and 100 vacuum tubes. Due to a lack of space at Birkbeck the calculator was built in Welwyn Garden City at the premises of the British Rubber Producers' Research Association (BRPRA), who sponsored the project.

In 1946 Bernal obtained funding from the Rockefeller Foundation for Andrew Booth to visit US researchers working on computers. Andrew Booth reported that only John von Neumann (a friend of Bernal's) at Princeton gave him any time. In 1947, with more funding from the Rockefeller Foundation, Andrew Booth undertook a six-month US tour based at the Institute of Advanced Studies at Princeton with John von Neumann. Booth was accompanied by his research assistant, Kathleen H V Britten, who was soon to become his wife.

It is difficult for the modern reader to appreciate the challenge faced by the computer pioneers, lacking any 'blueprint' for the architecture or components of

Andrew Booth Andrew Donald Booth was born in Weybridge, Surrey, in 1918. After a spell as a mathematics undergraduate at Cambridge in 1937/8, Andrew studied for a London External Degree whilst working in industry. During the war, one of his jobs was at an aero-engine factory in Coventry. Here he set up an X-ray department for the inspection of engine components. A Ph.D. from Birmingham University followed. In 1946 Booth joined the Physics Department at Birkbeck College, London, where he built several successively more powerful digital computers over the next ten years. He was a member of the first Council of the British Computer Society when it was formed in June 1957. In the same year he initiated an M.Sc. course in Numerical Automation at Birkbeck, one of the earliest degree courses in computing. Booth moved to Canada in 1962, retired in 1978 and died in 2009.

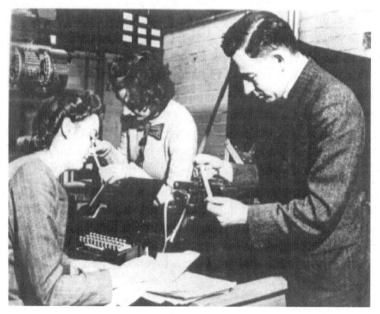

The Automatic Relay Calculator (ARC) with its inventor, Andrew Booth (right), Kathleen Britten (later Booth) on the left, and Xenia Sweeting in the background, in late 1946

a computer. By their discussions with John von Neumann, Andrew and Kathleen Booth were introduced to what we now regard as the standard structure of a computer, which is today widely known as the 'von Neumann architecture'. This follows von Neumann's 1945 *EDVAC Report*, as described in Chapter 1. Inspired by this discussion, Andrew Booth redesigned ARC to have a von Neumann (or EDVAC) architecture. This machine, ARC2, used 800 high-speed relays to form a parallel-operation single address code computer.

THE CHALLENGE OF MEMORY

The heart of the von Neumann architecture was the memory. Andrew and Kathleen Booth set out the technological options for the components of a computer with a von Neumann architecture in a paper that circulated among the growing community of computer pioneers during 1947. Such was the interest in it that they produced a second edition later that year. In their paper the Booths evaluated all of the physical properties that could be used for storage, or memory, including heat, light, sound and magnetism, and concluded that magnetism offered the best prospects because of its persistence.

Andrew Booth was interested in building a low-cost computer, and so he needed low-cost components. On his trip around the USA he had seen a simple recording device, sold for use in commercial offices, which allowed managers to record letters on to magnetic oxide coated paper discs for subsequent typing by their secretarial staff. However, in order to achieve the performance needed for it to act as the memory of a computer, he had to rotate the paper disc much faster than for simple voice recording. At this higher speed it proved impossible to keep the disc flat, and so he had to abandon this first attempt at a floppy disc.

Undaunted, Booth decided to try a different approach, and designed a memory using a brass drum with a nickel coating around the outside. The first drum was mounted on a horizontal axle and was about the size of a cotton reel, being 2 inches in diameter, with a modest packing density of just 10 bits per inch. Thus it was that he built the world's first rotating electronic storage device – albeit a drum rather than the now ubiquitous disc. The drums were built by his father, and together they created a company called Wharf Engineering Ltd, which manufactured drums and other computer peripherals. The **prototype magnetic drum** is now on display in the Science Museum, London.

Having proved that the drum concept worked, the Booths built a larger nickel-plated brass drum with 256 words of 21 bits to provide the memory for ARC2. It was completed by the end of 1947 and became operational, and it was still in service at Birkbeck in 1954. Its design was given to Adriaan van Wijngarden in Amsterdam for the construction of ARRA, the first computer in the Netherlands.

During his 1947 visit to America, Andrew Booth met Warren Weaver, Natural Sciences Division Director of the Rockefeller Foundation, who had funded the trip. Booth asked whether the Foundation would fund an electronic computer for London University. Weaver said that it could not fund a computer for mathematical calculations but that he had begun to think about using a computer to carry out natural language translation, and that the Foundation could fund a computer for research in that area.

Andrew Booth's prototype magnetic drum of 1947

As a result, Birkbeck became for the next 15 years a leading centre for natural language research. Initially the tiny memory on computers meant that it was very difficult to do any serious natural language processing. However, the Booths with their research students developed techniques for parsing text and also for building dictionaries. They published numerous books and papers on text processing, including creating Braille output and natural language translation. On 11 November 1955 the laboratory gave an early public demonstration of natural language machine translation.

COMPUTERS FOR ALL!

Even in those days of cumbersome early machines Andrew Booth wrote about making computers available as widely as possible. His stated ambition was to build computers sufficiently cheaply that every university could have one. This was radical thinking when Alan Turing's boss, the Director of NPL, was claiming that the ACE would be sufficient for the computing needs of the whole of the UK!

Even so, in 1949, under the unpromising heading of 'Desk Calculating Machines', Andrew Booth started a project that now seems well ahead of its time. A report from 1950 by Booth has recently been found in the Science Museum's archives. In it he evaluates the technical options for putting computers on, if not the desktop, at least the laboratory bench. The design used dekatron tubes, which operated on a decimal basis and thus provided simple counting devices. He lacked any simple way to display digits electronically and proposed using the position of the dekatron's lighted cathode as a counter, to be read in a manner similar to reading an analogue clockface. He observes in the report that this feature had (probably correctly) been regarded as a serious shortcoming by reviewers of his 1949 project from the National Physical Laboratory. However, he predicts correctly that a technical solution would soon become available. The project appears to have ended prematurely without a full prototype being built.

During 1948 Andrew Booth redesigned the ARC2 as an entirely electronic machine, which he called the **Simple Electronic Computer (SEC)**. This was built by Norman Kitz (formerly Norbert Kitz) during 1950.

An interesting historical footnote is that Norman Kitz left Birkbeck to work for English Electric at NPL on the DEUCE computer. From there he moved to Bell Punch and designed the world's first electronic desktop calculator, called ANITA. So although Andrew Booth never completed a desktop calculator at Birkbeck, it seems likely that he inspired one of his students to do so.

The SEC (Simple Electronic Computer) being tested by Norman Kitz in 1950

SEC used a comparatively small number of vacuum tubes (230) in its construction and employed a two-address set of operation codes. Although Norman Kitz completed its construction, the machine never became operational. Using the lessons learned from SEC, Andrew Booth moved swiftly on to create his best computer design, the **All-Purpose Electronic Computers**, (which were known as APE(X)C, where the X is the initial of the sponsoring agency).

The first machine, known as APE(R)C, first ran successfully on 2 May 1952. The (R) signified the British Rayon Research Association (BRRA), for whom it was built. It was constructed in a barn at the Booths' home in Warwickshire, before being moved to Birkbeck for commissioning and subsequent delivery to the BRRA in Manchester in July 1953.

Booth's All-Purpose Electronic Computer, known as APE(X)C and put to use for X-ray crystallography at Birkbeck College, in 1953

Once the APE(R)C computer was moved to Birkbeck, construction of a larger version, known as APE(X)C (the X standing for X-ray), for the Birkbeck X-ray crystallographers began. APE(R)C used 500 vacuum tubes and had a drum with 512 words of 32 bits rotating at 3,800 rpm; it ran at 30 KHz. By removing some redundant logic, APE(X)C was made to use only 320 vacuum tubes, but it ran at 60 KHz and had a much larger drum with over 8,000 words of 32 bits. Input–output on APE(X)C was by either teletype or punched cards.

THE BOOTH MULTIPLIER

If the drum reflected Andrew Booth's engineering talent, then the **Booth multiplier** was a demonstration of his mathematical skill. A key component of any computer design is the arithmetic unit, and to provide fast arithmetic it is necessary to have hardware multiplication and division. When the Booths visited von Neumann in 1947 they obtained details of his design for both a hardware multiplier and a divider. Andrew Booth described the latter in a later interview as 'a beautiful divider' but the multiplier as 'an abortion'. When he asked von Neumann why he had not used a similar approach in his multiplier as in the divider von Neumann assured him it was a theoretical impossibility, and Booth accepted the great man's

The Booth multiplier The Booth multiplier follows the usual method for decimal long multiplication of summing partial products. However, it also uses a 'trick': to multiply by a string of 9s it is possible to shift left by an appropriate number of places and subtract the multiplier from the result. This approach works even better in binary, where it results in a simple rule. The procedure is as follows.

1. Examine each pair of digits in the multiplier, starting with the least significant and creating the first pair by appending a dummy zero at the least-significant end. Then:
   - if the pair is 01, add the multi-plicand;
   - if the pair is 10, subtract the multiplicand;
   - otherwise, do nothing.
2. Shift the partial product one place right and examine the next pair of digits.
3. Repeat as many times as there are digits in the multiplier.

opinion. He recalled that when he was designing the APE(R)C computer he realised that von Neumann was wrong, and recollected how in 1950, over tea with his wife in a central London cafe, he designed a binary multiplier, which, with subsequent minor modifications, is the Booth multiplier that is still in use in some computers today.

COMMERCIAL SUCCESS

Accommodation at bomb-damaged Birkbeck College in central London was at a premium throughout this period. This is why the Booths built their computers in a barn at their home in Fenny Compton, Warwickshire, where Andrew Booth's father also lived.

It was to this barn in a freezing March 1951 that a three-man team led by Raymond 'Dickie' Bird from British Tabulating Machines (BTM) came to visit. BTM manufactured electromechanical equipment for business data processing, based on the Hollerith type of punched cards. BTM were the UK's leading supplier of punched-card systems, and their management had decided that they needed a small electronic computer to improve the calculating power and flexibility offered by their tabulators.

At the time when BTM joined forces with Andrew Booth there were, as described elsewhere in this book, several other electronic computer projects in the UK. Strong links had already developed between the EDSAC team at Cambridge University and J Lyons & Co. Ltd, who were building their LEO (Lyons Electronic Office) computer. Manchester University was forging links with Ferranti Ltd, while NPL with Pilot ACE had joined forces with English Electric. Further, the latter two were building large and relatively expensive scientific computers. Elliott was ploughing its own furrow and soon would have the help of the National Research Development Corporation. In 1951 BTM appeared to be losing out and consequently had limited options.

In just a few days Raymond Bird's team had copied Andrew Booth's circuitry from APE(R)C. Returning to BTM's factory at Letchworth, they added extra input–output interfaces and named the resulting machine the **Hollerith Electronic Computer (HEC)**. This prototype computer, HEC1, still survives, unlike so many early machines that were dismantled when no longer needed, and is now in store in the Birmingham Museum. It is one of the world's earliest surviving electronic computers.

BTM moved ahead rapidly, getting HEC1 to work by the end of 1951. BTM management decreed that the HEC would go to the Business Efficiency Exhibition in October 1953, and so a new machine, **HEC2**, had to be built. It was contained in a smart metal cabinet suitable for the public to see. Two pieces of software were written to attract public interest at the exhibition: one read in 13 out of 52 specially punched cards, each representing a different playing card, and proposed an opening contract bridge bid; the other played noughts and crosses.

Starting in 1955, seven similar machines were built under the name of the HEC2M, mainly for technical applications. The selling price was about £20,000. The successor to the HEC2M was the HEC4, which was for commercial data processing, with about 100 being sold in the UK and abroad. The HEC4 ran at 30 KHz and initially had a drum memory with 1,024 words of 40 bits, rotating at 3,000 rpm. The processor cabinet was just over 2 m long, 0.6 m wide and 2 m high and contained 1,100 vacuum tubes.

With the HEC4 BTM had entered the data-processing market with a straightforward, reliable machine at modest cost that suited their existing large customer base. At the end of the 1950s HEC4 was the UK's best-selling computer by volume. BTM renamed the HEC4 the BTM 1201 and subsequently doubled the drum size in the BTM 1202. After BTM merged with Powers SAMAS in 1958–9 to form

The HEC1 (Hollerith Electronic Computer), built by the British Tabulating Machines Company (BTM) to Booth's design, in 1951 (the original can still be seen at the Birmingham Museum).

An HEC2 computer (its black cabinets can be seen in the background) playing noughts and crosses at the Business Efficiency Exhibition in London in October 1953

International Computers and Tabulators Ltd (ICT), these machines became the ICT 1200 range.

Andrew Booth continued to build new machines. After APE(X)C came MAC (Magnetic Automatic Calculator). Three examples of a development of MAC named M.2 were built by Wharf Engineering Ltd for University College London, Kings College London and Imperial College London. The computer supplied to UCL was also used to teach the first programming courses given there. The keynote of the M.2 was, as in previous machines, its small size and simplicity. Of the M.2 it was said that it

> occupies a space rather less than that of an office desk, consumes as much power as an electric fire, but has roughly the speed and capacity of the much larger commercial machines which are being provided for some of the smaller Universities.

One notable landmark was Kathleen Booth's book on programming the APE(X)C computer. Published in 1958, this was among the early books on programming and was unusual in having a female author. Kathleen did most of the programming while Andrew Booth concentrated on the design of their computers.

Andrew and Kathleen Booth resigned from Birkbeck at the end of the 1961–2 academic year. They moved to Canada, where Andrew continued his distinguished academic career – initially at the University of Saskatchewan and subsequently as President of Lakehead University, Ontario.

7
INTO THE MARKETPLACE

Simon Lavington

OUT OF THE LABORATORY

The preceding chapters have described the main British research projects that were active from 1945 to 1952. What was it that pulled these pioneering digital designs out of the laboratories and into the real world? The answers are to be found by looking at the needs of real-world users with real-world problems. First, though, a word of caution: different users had different traditional ways of solving problems and different views of whether the new breed of stored-program digital computer could be of any help. We shall deal with three applications areas, to make the point that there were interesting variations in the rate at which different groups of people began to take an interest in the new inventions emerging from the laboratories. The three areas are:

- defence;
- science and engineering;
- commercial business data processing.

DEFENCE AND THE COLD WAR

With the Soviet blockade of Berlin in 1948 and the Korean War (1950–3), there arose an urgent need for more powerful computers in at least four areas of defence:

- signal interception (code-breaking);
- ballistics and gunnery control;
- radar signal analysis;
- simulations of large and complex systems such as nuclear reactors and guided missiles.

The use of digital electronics for code-breaking had, of course, been thriving in secret at Bletchley Park since 1943, when the Colossus series of so-called Rapid Analytical Machines began work. Although they were not true general-purpose stored-program computers, the Colossus machines did create a climate within which forward-looking GCHQ people started to consider special-purpose digital computers in the early 1950s. The secret OEDIPUS machine, operational in 1954, was one such. It had a magnetic drum store developed by Ferranti Ltd and a semiconductor associative (i.e., *content-addressable*) store developed by Elliott Brothers (London) Ltd. GCHQ also commissioned the Elliott 153 general-purpose digital computer for rapid DF (direction finding) of potentially hostile signals.

In the areas of ballistics, gunnery control and radar signal analysis the government financed the construction of several general-purpose digital computers. These, along with the Elliott 153, are listed for completeness in **Table 7.1**. Whilst all of these machines no doubt gave valuable service in the defence of the UK during the first decade of the Cold War, they were all classified, and few of their designs had much impact on the civil arena.

Name of computer	Where designed and built	Initial target applications	Date first working	Customer	Location
152	Elliott's Borehamwood Labs	Naval gunnery (anti-aircraft fire control)	1950	Admiralty	Computing Division, Borehamwood
Nicholas	Elliott's Borehamwood Labs	Trajectory calculations for a guided bomb	1952	Royal Aircraft Establishment (RAE), Farnborough	Theory Division, Borehamwood
153	Elliott's Borehamwood Labs	Rapid plotting of DF (direction finding) intercepts	1954	Admiralty and GCHQ	Irton Moor, Scarborough
403	Elliott's Borehamwood Labs	Analysis of Woomera's test-firing data	1956	Long Range Weapons Establishment	Salisbury, South Australia
TREAC	Telecommunications Research Establishment (TRE)	In-house mathematical support	1953	TRE	TRE, Malvern
MOSAIC	Post Office Research Station, Dollis Hill	Analysis of radar tracking data	1953	Ministry of Supply (MOS)	Radar Research and Development Establishment (RRDE), Malvern

Table 7.1. Six post-war defence-related computer projects

Four of the secret projects listed in Table 7.1 have already been mentioned in Chapter 5, since they originated in the Borehamwood Laboratory of Elliott Brothers (London) Ltd. MOSAIC was inspired by Alan Turing's ACE project, as mentioned in Chapter 2. TREAC used a similar technology to that of the Manchester University projects described in Chapter 4 except that it was bit-parallel, not bit-serial. There may have been other secret projects. If there were others, they would probably have been for special-purpose (like OEDIPUS) rather than general-purpose computers.

When considering the simulations of large and complex systems such as nuclear reactors and guided missiles another significant factor comes into the picture. Digital computers were not at first usually fast enough to be cost-effective. Large engineering simulations remained the province of analogue computers until at least the end of the 1950s. The preference for analogue computers was also seen in military aircraft, where reliability and compactness were as vital as speed. *Flight control* (e.g. stabilisation) remained the province of analogue computers until about 1975, though airborne digital computers began to be introduced for *mission systems* (e.g. navigation and weapons-aiming) from the late 1960s. In summary, it was many years before digital computers became the preferred solution to a number of specialist problems.

SCIENCE AND ENGINEERING

Since the pioneering laboratories at NPL, Cambridge, Manchester, Borehamwood and Birkbeck were by their nature staffed by mathematicians and engineers, it was natural that their pioneering computers should at first be oriented towards the needs of science and engineering. In these areas there was a tradition of numerical computation. The diagram overleaf is a reminder of how **research projects at five laboratories** led to liaisons with manufacturers who produced the first digital computers to arrive in the marketplace.

Whilst all of those developments were matched by other and grander projects in America, it is clear that by 1952 a small but thriving independent British computer industry was firmly in place. However, with the exception of LEO and the later HEC4 computers all the early machines were for initial applications in science and engineering. For these areas digital computers replaced the old hand-operated desk-top mechanical calculators. Some idea of the advantages offered by the new electronic machines may be had from the following quotation, taken from a 1953 brochure for the Elliott 401 computer.

> For example, the [Elliott] computer will determine all the roots
> of a set of 20 simultaneous linear equations in about three minutes
> or all the roots of a set of 30 equations in about 20 minutes. Using
> desk machines the determination of a dominant root of a set of 20
> equations requires about 10 days work and a set of 30 equations
> is liable to be too formidable a problem to be tackled on desk machines.

It is not surprising that the first customers of the computers built by
Ferranti Ltd, Elliott Brothers (London) Ltd, English Electric and the
British Tabulating Machine Co. Ltd were in science and engineering –
as shown in **Table 7.2**.

There is one entry in Table 7.2, namely the 1953 delivery of a Ferranti
Mark I*, which deserves further comment. At the time the final destination of this computer was classified information. On the day of
delivery the drivers of two large Ferranti lorries containing the computer were simply instructed to park their vehicles in a designated
lay-by, leave the keys and walk away. The lorries were then taken over
by people from the Ministry and driven to a secret location – disclosed
years later as being GCHQ at Cheltenham. This is believed to have
been the first general-purpose computer to be acquired by the code-
breakers at GCHQ.

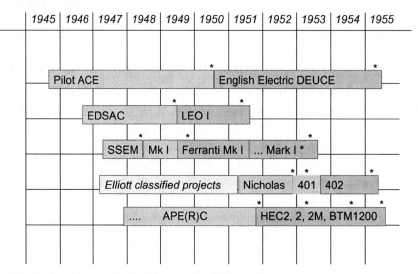

* Indicates date on which each computer first ran a program.

The research origins of the first British production computers

Year delivered	Computer	Customer	Application
1951	Ferranti Mark I	Manchester University	Scientific and engineering
1952	Ferranti Mark I	Toronto University	Mathematical research
1953	Ferranti Mark I*	Ministry of Supply (GCHQ, Cheltenham)	Classified work
1954	Elliott 401	Agricultural Research Council, Rothamsted	Agricultural statistics
1954	Ferranti Mark I*	Royal Dutch Shell Labs, Amsterdam	Oil refining studies
1954	Ferranti Mark I*	Atomic Weapons Research Establishment, Aldermaston	Research work
1954	Ferranti Mark I*	A V Roe and Co. Ltd, Manchester	Aircraft design calculations
1955	Elliott 402	Institut Blaise Pascal, France	Mathematical research
1955	Elliott 402	Army Operational Research Group, West Byfleet	Operational research
1955	BTM 1200 (HEC2M)	GEC Research Labs, Wembley	?? (application unknown)
1955	BTM 1200 (HEC2M)	ESSO Oil Refinery, Fawley	Scheduling and planning
1955	Ferranti Mark I*	National Institute for Applications of Mathematics, Rome	Research work
1955	Ferranti Mark I*	Ministry of Supply, Fort Halstead	Defence-related research
1955	English Electric DEUCE	National Physical Laboratory, Teddington	Mathematical applications
1955	English Electric DEUCE	Royal Aircraft Establishment, Farnborough	Aircraft research

Table 7.2. The first five years of UK computer production: total deliveries to external customers, 1951–5

Before leaving science and engineering, we should mention three smaller British computer projects that used electromechanical *relays* for their internal logic elements. A relay is an electrically operated switch whose contacts can be closed or opened by means of small electromagnets. The switching time of each relay was at least 1,000 times slower than that of a vacuum tube, and so arithmetic units based on relays were very much slower that those based on electronics. Historically speaking, relay-based computers overlap with the earlier form of *sequence-controlled calculators*, which generally held their instructions on some form of semi-permanent external storage such as punched paper tape. The three machines described below normally operated in the sequence-controlled mode.

The largest, though not the fastest, of the British post-war relay computers was called ICCE – Imperial College Computing Engine. As its name implies, it was built in the Department of Mathematics at Imperial College, London. ICCE was working by about 1952.

A decimal computer using a mixture of relays and electronics was designed and built at the Atomic Energy Research Establishment, Harwell. It was first working in 1951 and operated at Harwell until 1956. In 2008 it was loaned to the National Museum of Computing at Bletchley Park, where, at the time of writing, it is being restored to working order.

The design of the third relay computer was begun at the Royal Aircraft Establishment (RAE), Farnborough, and was called RASCAL (RAE Sequence Calculator). It was still under construction in 1953 and seems never to have been completed. It was superseded at RAE by the arrival there in 1955 of a true stored-program computer, an English Electric DEUCE.

THE WORLD OF COMMERCE AND BUSINESS

The idea that digital computers could be used to handle currency, names and addresses, product expiry dates and all the other alphanumeric information of the business world was slow to catch on. Apart from cost and reliability considerations, the early pioneering computers simply lacked the facilities for handling large volumes of input and output data. The LEO computer, a direct descendant of the Cambridge EDSAC, was the first machine to show that commercial data processing could indeed be cost-effective. This was followed in the UK by BTM's HEC4 and Elliott's 405.

The commercial take-up was still slow compared with that in science and engineering. The British bank that was probably the first to install its own computer did not do so until early 1961, as described below. By the end of 1959 there were about 170 computers installed and working in the UK. All but ten were of British manufacture. Of the 170, about 48 per cent were doing scientific and engineering tasks, 40 per cent were applied to business and commerce and the remaining 12 per cent were installed in computing service bureaux and training centres. By 1959 International Computers and Tabulators (ICT, the successor to BTM) was making almost half of the computers being applied to businesses.

If 170 seems today like a very small total for the number of computers installed in Britain, it certainly exceeded the expectations of earlier experts. In September 1951 Professor Douglas Hartree had remarked:

> We have a computer here at Cambridge; there is one at Manchester and one at the NPL. I suppose there ought to be one in Scotland, but that's about all.

The uncertainty of the times is reflected in his explanation that one of his reasons for nominating Scotland was '... to disperse the available equipment in case of war'. Hartree, of course, had been a central figure in organising wartime computing activity, and his estimate of only three or four high-speed digital machines reflected his memory of wartime defence requirements such as the solving of differential equations. Few of the early pundits thought in terms of business and commerce.

The reluctance to move to electronic computers is well illustrated by the case of Martins Bank. At the end of the 1950s the bank began experimenting with a Pegasus computer at Ferranti's London Computer Centre. A current-account bookkeeping program was developed there and went live in January 1960. As a result of this experience, a Ferranti Pegasus II computer was eventually installed at Martins Bank's Liverpool headquarters in April–May 1961. Computer accounting was first reliably operational for a Liverpool branch of the bank by the end of 1961.

This happened at a time when people were afraid that 'computers will take over our jobs'. An August 1961 press release from Martins Bank, announcing the successful computerisation of customers' current accounts, included a paragraph headed 'No Redundancy'. This declared that:

> The purpose of the computer is to reduce the amount of manual work needed in bank work and to relieve staff of the monotony that in the past has been characteristic of much of bank routine. Nevertheless redundancy is not expected. The natural wastage of staff performing routine work will permit numbers employed to be reduced if necessary. However, experience indicates that the growth of business and extension into new fields tends to outweigh economies in staff effort due to improved systems with a result that staff numbers do not decrease, but instead the same size of staff is needed for the greater amount of work to be done. As a result the tasks to be performed are more varied, and the opportunities for staff advancement considerably increased.

Business may indeed have grown, and the tasks carried out by most staff probably did become more interesting. Nevertheless, for reasons unconnected with computers, by the end of 1969 Martins Bank had vanished from sight in a merger with Barclays Bank.

THE MARKET GROWS AND THE MANUFACTURERS SHRINK

By 1962 the rate of the 'computerisation' of the UK had accelerated. In June 1962 there was a total of about 380 computer installations in the country, supplied by 14 manufacturers (eight British, five American and one French). The machines ranged in power from *supercomputers*, each occupying one or more very large rooms, to modest installations, each looking like half a dozen filing cabinets and filling a medium-sized office. There was no internet, no email and normally no communications link between one computer and another. If data had to be transferred, it was usually a matter of physically carrying a reel of magnetic tape between one installation and another.

British manufacturers were slow to develop suitable magnetic tape systems, and IBM's American-designed tape equipment and magnetic recording standards soon became the preferred means for information interchange worldwide. The phrase 'IBM-compatible' brought a warm feeling to many first-time computer users. Significantly, a single American company, IBM, accounted for almost 20 per cent of the installed computers in the UK by June 1962. The golden age of home-grown British computers was starting to come to an end.

Within the UK, market forces and the increased penetration of American suppliers were obliging the British manufacturers to rationalise. Between 1959 and 1968 there was a flurry of mergers and takeovers, culminating in **the formation of International Computers Ltd (ICL)**. With encouragement and financial backing from the government, this company became the country's single major home-grown supplier of mainstream computers. It is important to use the word 'mainstream' here because ICL chose not to cover areas such as **industrial process control** and online real-time defence applications. Digital computers for these areas continued to be designed and built by Ferranti Ltd and by GEC-Marconi, the successors to Elliott-Automation.

All of the company names associated with the early British computers have now vanished from the public gaze. ICL was gradually absorbed into Fujitsu during the period from 1981 to 2001. In the defence and aerospace area, the current UK successor to Elliott-Automation, GEC, Marconi and Ferranti is in effect BAE Systems, a large company very much alive at the time of writing.

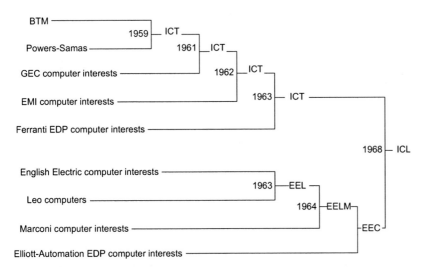

The formation of International Computers Ltd (ICL) as a result of the gradual coming-together of all the early British computer manufacturers

Computers for industrial process control continued to be built by companies other than ICL. This photograph shows an Elliott computer installed in 1967 at Dista Products Ltd, Liverpool, for the control of batch fermentation of antibiotics such as penicillin. The complete Elliott-Automation control system cost about £100,000 (1967 prices), of which the digital computer at its heart represented only about 20 per cent. The benefit to Dista Products was greater accuracy in controlling parameters such as temperature, giving higher yields of antibiotics. From 1958 onwards, manufacturers of a variety of products and services installed process-control computers as industrial automation started to take off.

HINDSIGHT AND FORESIGHT: THE LEGACY OF TURING AND HIS CONTEMPORARIES

Simon Lavington

WHO DID WHAT, AND WHEN?

It might be said that the theoretical basis for all modern computers stems from Alan Turing's 1936–7 paper 'On Computable Numbers', and that the physical structure of all computers stems from John von Neumann's 1945 *EDVAC Report*. This is not quite the full picture. However, trying to be precise about historical connections seems rather academic because computing has changed in such dramatic ways since the early 1950s. In any case, no single individual or laboratory was indispensable to the birth of the Information Age.

In the context of early British computers, it is nevertheless interesting to summarise the material in earlier chapters by linking the British pioneers with some practical innovations that have stood the test of time. Leaving Turing aside for one moment, here are some examples of why users of modern computers might be thankful that particular individuals did what they did in the period 1948–54.

- *Maurice Wilkes and colleagues at Cambridge.* This group pioneered the idea of assemblers and a user-friendly computing service. By means of seminars, books and short courses, Cambridge spread information widely and encouraged cross-fertilisation between other laboratories.
- *Freddie Williams, Tom Kilburn and colleagues at Manchester.* This group pioneered ideas about memory management and the addressing of structured data. The group was also at the forefront of developing autocodes and early high-level languages.
- *Bill Elliott, Andrew St Johnston and colleagues at Borehamwood.* This group pioneered modular circuit packaging and the production of reliable, affordable computers.

- *Andrew Booth at Birkbeck.* He also championed the idea of small, affordable computers. He was the first to suggest a magnetic drum as a cost-effective secondary store, and set BTM on the path of electronic digital computers.

All those listed above were relatively young people. In 1948 St Johnston was the youngest at 26, with Williams the oldest at 37, so of course the careers of all these pioneers continued into the 1970s and (except in the case of Williams, who died in 1977) many years beyond. In contrast, Alan Turing, though he was a year younger than Williams, died in 1954 aged only 42.

How did his computing contemporaries regard Turing? What influence did Turing have upon them, and upon the course of computer development, in the immediate post-war years? These are difficult questions to answer with precision, even with the benefit of hindsight. It seems helpful to attempt a description of Turing's influence in two periods: first during his lifetime and then 30 years after his death.

TURING AS SEEN BY HIS CONTEMPORARIES

To his contemporaries in the small world of early British computing, **Alan Turing** was certainly well known – even held in awe. If his ideas on computer design were not as influential as might be supposed from his renown, this is possibly because he was not by nature an easy person with whom to communicate. He would often resort to deriving solutions rapidly from first principles, a tendency that caused Douglas Hartree to remark rather unfairly that Turing had 'a unique talent for making even the simplest things look complicated'. His thoughts were also likely to be racing ahead, making him sometimes intolerant of interruptions or questions. At NPL in 1946–7 his many revisions of his design for the ACE computer must have exasperated his colleagues. In his history of NPL David Yates remarks:

Alan Turing in 1951 in a portrait photograph taken upon his election as a Fellow of the Royal Society. © Godfrey Argent Studio.

> Turing's combination of dominance of the project with a lack of ability to collaborate, and a lack of interest in the organisation needed to get practical development moving, must have constituted a

Two curious aspects of ACE Of the many interesting features in Alan Turing's original design for ACE, two may seem strange to modern computer experts. Firstly, an ACE instruction made no explicit mention of 'op codes' or a main 'accumulator'. On a modern computer, the accumulator is a special register or short-term store whose primary job is to hold the number resulting from each individual instruction immediately after it has been obeyed. The op code specifies the function, for example ADD or SUBTRACT, that an individual instruction performs.

Instead of this, each ACE instruction gave a source number (S) and a destination number (D), these numbers being used for a variety of purposes. Some of the ACE source and destination numbers specified one-word registers called 'temporary stores' (TS); other numbers specified main memory addresses. Operations in ACE were implied by yet more source/destination identification numbers. Thus, to perform a logical AND operation one simply specified a source number 25. This had the effect of performing the logical AND operation on the bit patterns contained in temporary stores TS14 and TS15. For example, an ACE instruction of the form

`S = 25, D = 13`

resulted in the following operation when expressed in a modern notation:

`TS13 := TS14 & TS15.`

A look at the layout of the DEUCE instruction (see Appendix A), which was similar to that of the Pilot ACE, will help to explain this complex use of source and destination registers.

significant factor in the delay in implementing his plans.

It seems that Turing was not really a team player at NPL.

Why didn't Turing freeze his ACE design in 1947 and publish the ideas more widely? Andrew Hodges, his biographer, has remarked that:

> If in 1948 he had written a serious monograph on 'The theory and practice of computation', making it clear how the computer was based on the Universal Turing Machine of 1936, explaining the significance of that universality, with the unlimited prospects for programming languages and the connection with symbolic logic, together with an analysis of the physical basis of storage mechanisms etc., it would have set an unassailable world lead. But he did nothing of the kind.

Even if Turing had published more widely, it is far from certain that his architecture and register-level structure for ACE would have been adopted by other computer designers. His pursuit of hardware cost-effectiveness imposed a considerable burden on the programmer, and **ACE's design had some curious aspects**. Other designs may not have been so efficient, but they were generally much more user friendly. And yet, looking at his earliest ACE report, we can appreciate his understanding of many end-user concepts that are now taken for granted. Amongst these are his thoughts on the careful preparation of programs and the use of subroutine libraries.

To the modern enquirer one of the most puzzling questions concerns Cambridge University. Turing was clearly content with life as a Fellow of King's College, often returning there for a short holiday during the last six summers of his life. Why didn't he take more interest in the design of the Cambridge EDSAC? Whilst on sabbatical from NPL from September 1947 to September 1948 he chose to spend most of his time at Cambridge. Yet it was not until May 1948 that Alan Turing finally visited Maurice Wilkes at the Computer Laboratory, where the implementation of EDSAC was making good progress. Turing was later to remark of this meeting: 'I couldn't listen to a word he said'.

Although Wilkes and Turing had been mathematics undergraduates at the same time many years previously, their mathematical interests were always poles apart. By 1948 their personalities and ambitions had also become irreconcilable. As early as December 1946, when Turing first learned of the detailed EDSAC plans, he wrote:

> The 'code' which he [Wilkes] suggests is however very contrary to the line of developments here [at NPL], and much more in the American tradition of solving one's difficulties by means of much equipment rather than by thought.

In turn, Wilkes wrote that he

> found Turing very opinionated and considered that his ideas were widely at variance with what the main stream of computer development was going to be.

Of all the machines described in this book, perhaps the Elliott 153 computer is the only one to come close in spirit, if not in detail, to Turing's ideal of trying to maximise the speed and functionality to be derived from bit-serial hardware. The Elliott designers approached this ideal by arranging low-level functional parallelism with multiple data highways and by choosing a relatively long 64-bit instruction. The Elliott 153 computer was designed for a particular defence application where cost was not a prime consideration. Turing worked with much less hardware and a shorter but much more complex instruction format.

Of the machines mentioned in this book that were intended for the open market, the DEUCE obviously followed the Pilot ACE design because of the links between NPL and English Electric. The Elliott 402 and the BTM HEC shared one aspect with the Pilot ACE, in that their instruction formats permitted the address of the next instruction to be specified. This was because the main store for both the 402 and the HEC consisted of a relatively slow drum. In all other respects the architectures of those two machines were quite unlike that of the Pilot ACE. Indeed Andrew Booth, the inspiration behind the HEC computer, said of the ACE instruction set that 'the code is rather complex, and more suited to the needs of

Ace and conditional branching.
Another strange aspect was the absence in the original 1945–6 ACE specification of any explicit instructions for conditional branching. A conditional branch instruction on a modern computer typically tests the value held in the accumulator and performs one of two actions depending on whether the test 'passes' or 'fails'. For example, to test whether a number is negative (i.e., less than zero) the computer might perform the equivalent of:

```
If [contents of acc] < 0
then go to the instruc-
tion labelled JA, other-
wise go to JB.
```

All computers, including ACE, allocate one bit in a word to represent the sign of a number. Let us call this bit D. When performing a conditional branch, we are in effect asking the computer to choose from the alternative destination instructions at addresses JA and JB, according to whether a certain digit D is 0 or 1. This is equivalent to saying:

```
If D = 1 then go to JA,
else if D = 0 then go
to JB.
```

In the original ACE that action could not be expressed as a single explicit instruction. It was first necessary to do the following computation on the 'next instruction' part of an instruction B held in the store.

```
('next instruction' part
of B) := ((D × JA) +
(1 - D) × JB).
```

We then cause the new instruction B to be obeyed.

mathematical logic than of the processes of arithmetic'. Returning to Turing's idea of specifying the next instruction address and the consequential discipline of 'optimum programming', these ideas did not make sense if the primary memory was random access, as in the Manchester computers. Once cost-effective RAM became more widely available in the late 1950s Turing's ACE philosophy became somewhat redundant.

What of Turing's contemporary influence as a *user* of computers? One has to admit that he mostly ploughed a lonely furrow. After writing the first *Programmers' Handbook* for the Ferranti Mark I computer in March 1951, a complicated 110-page document, all of his subsequent programming activity at Manchester seems to have been directed towards his own private research into morphogenesis. Of course he also had a long-standing interest in what would now be called artificial intelligence, but there is little evidence that he wrote any significant programs that explored this subject. For the last two years of his life Alan Turing appears to have taken no interest in the hardware or software, or indeed the applications, of the many computer projects that were coming to fruition in Britain and America in the early 1950s. Perhaps, having tried to promote his own ideas on computer design in the period 1945 to 1947 and then seen these ideas largely ignored outside NPL, he had decided to move on to other, more challenging, things. Certainly morphogenesis was, and still is, a challenging subject.

TURING'S REPUTATION BY 1984

Thirty years after his death, had Alan Turing's influence and reputation grown or declined? In very many ways it had grown. By 1984 both artificial intelligence and computational theory had for some years been well-established sectors of the overall discipline of computer science. In both these sectors he was regarded by many as the founding father. By 1984, too, western society was becoming liberated from the sexual constraints of the 1950s. In 1983 Andrew Hodges had published his authoritative biography – the first deep study of Turing's thinking and motivation. Alan Turing's persecution by society for his homosexuality in the period leading up to his tragic death gradually became a cause célèbre. In 2009 Gordon Brown, the British Prime Minister, publicly apologised for society's earlier treatment of Turing.

Perhaps of more significance for Turing's posthumous reputation, in the early 1970s details of the wartime cryptanalysis efforts at Bletchley Park started to be released into the public domain. It then became possible for people to understand why he had been awarded an OBE at the end of the war. Whilst certainly not the only brilliant mind to work at

Bletchley Park, he was undoubtedly held in high esteem by colleagues, who referred to him as 'the Prof' – the person to whom you went if you had a seemingly intractable problem. For his work there he had by the 1970s become a 'national treasure' in the public's mind. So for Bletchley Park, and for several other reasons touched on in this book, Alan Turing surely deserves to be remembered today as an undoubted National Treasure.

APPENDIX A
TECHNICAL COMPARISON OF FIVE
EARLY BRITISH COMPUTERS

The sample electronic stored-program computers to be compared are:

	Manchester SSEM (the 'Baby')	Cambridge EDSAC	Ferranti Mark I	BTM HEC2M	Elliott 402	English Electric DEUCE
Refer to chapter	4	3	4	6	5	2
Date first working	June 1948	May 1949	Feb 1951	1955	1955	1955
Prototype of this computer	–	–	Manchester University Mark I	Andrew Booth's APE(X)C	Elliott 401	NPL Pilot ACE

The five early British computers that are compared in this appendix

Setting aside the SSEM, the other five together represent the different types of machine available for scientific and engineering applications at the start of the computer age. It is difficult to compare the end-user capabilities of these machines in an even-handed way because of the significant differences in their hardware architectures. The following table, which should be read in conjunction with its notes, gives a reasonable comparison of their basic technical characteristics.

	Manchester SSEM (the 'Baby')	Cambridge EDSAC	Ferranti Mark I	BTM HEC 2M	Elliott 402	English Electric DEUCE
Word length, bits [a]	32	35	40	32	32	32
Instruction length, bits	32	17	20	32	32	32
Instruction format	1-addr	1-addr	1-addr	1 + 1	1 + 1	2 + 1
Instruction set: number of ops [c]	7	18	47	16	16	Approx. 30
Primary store size [a]	32	512	512	1024	15	402
Primary store type [d]	CRT	Mercury DL	CRT	Drum	Nickel DL	Mercury DL
Secondary store size [a,b]	-	-	8K	-	3K	8K
Secondary store type	-	-	Drum	-	Drum	Drum
ADD time, min/max (millisecs)	1.2 / 1.2	1.4 / 1.4	1.2 / 1.2	1.25 / 21.25	0.204 / 13.204	0.064 / 1.064
MULTIPLY time, min/max (millisecs)	- / -	4.5 / 4.5	2.16 / 2.16	3.75 / 50.0	3.0 / 16	2 / 3
Digit period, microseconds	8.5	2	10	30	3	1
Main type of vacuum tube	EF50	EF54	EF50	6J6	12AT7	ECC91
Approx. number of vacuum tubes (incl. thermionic diodes)	563	3,000	3,600	700 ?	650 ?	1,450
Input medium	Switches	PTR	PTR	CDR	PTR	CDR
Output medium	Display CRT	PTP; printer	PTR; printer	CDP; printer	PTP; printer	CDP; printer
Approx. cost of a production model	-	-	£95,000	£20,000	£27,000	£45,000

Basic technical comparison of five early British computers

Notes

(a) Word length. In the case of EDSAC and the Ferranti Mark I, short words as well as long words could be addressed in memory. Therefore, memory capacity for these two machines is quoted in terms of short words. Many scientific programs would have computed using long words (35 bits for EDSAC, 40 bits for the Ferranti Mark I) as data. All computers in the Table except the SSEM preserved the double-length result of a multiplication. The Ferranti Mark I and the English Electric DEUCE also provided specific hardware support for double-length (respectively 80 bit and 64 bit) arithmetic.

(b) The figures show the drum capacity available for user programmers. In the case of the Ferranti Mark I there were up to 256 physical drum tracks, though only 64 of these were normally available. In theory the Ferranti Mark I could have had more drums attached, since a drum was synchronised to the CPU's clock. In the case of the BTM HEC2, Elliott 402 and English Electric DEUCE, the CPU was synchronised to a single drum.

(c) This is the number of effective distinct operations, as would be recognised by a modern programmer. The actual instruction layout in the case of the BTM HEC2, the Elliott 402 and especially the English Electric DEUCE would, however, look strange to modern eyes. Only the Ferranti Mark I and the Elliott 402 had explicit index registers (B registers) for address modification. An index register was added to EDSAC in 1955. For the English Electric DEUCE, a facility called Automatic Instruction Modification (AIM) was added in 1957 as an upgrade. None of the computers in the Table had floating-point hardware. An enhanced version of the Elliott 402, the 402 F, was produced in 1957 with floating-point hardware.

(d) Three of the computers in the Table had delay line primary stores, using either acoustic waves travelling in tubes filled with mercury or magneto-strictive waves travelling along nickel wire. The HEC2 had a drum primary store. The SSEM and the Ferranti Mark I used Williams–Kilburn CRT (electrostatic) storage. The choice of memory technology affected instruction times in the following manner. When a computer obeyed instructions directly from a delay line store or a drum store, access time was affected by the address of the instruction or operand relative to the current position of information circulating in the store. That is to say, access to primary memory was sequential and not (as in the case of a modern computer) random. Thus, minimum and maximum times are quoted in the Table for the ADD and MULTIPLY instructions. (Actually, the SSEM's only arithmetic operation was SUBTRACT but this does not affect the time comparisons). For computers with a (1 +1) or (2 + 1) address instruction format, optimum programming techniques could be used to try and keep execution times as close to the minimum as possible. In contrast, the SSEM's store and Ferranti Mark I's primary store was random-access, so optimum programming was unnecessary.

Below we give illustrative examples of the layout of instructions for four of the computers in the above tables. The first three machines all have the so-called 'one-address' instruction format, where each instruction specifies just one operand address at a time. The last example has a 'two-plus-one address' instruction format. Two of the addresses refer to operands; the extra 'plus one' address gives the location in store of the next instruction to be obeyed.

THE MANCHESTER SMALL-SCALE EXPERIMENTAL MACHINE (SSEM), KNOWN AS THE 'BABY'

The layout of a 32-bit SSEM instruction is shown in **the figure below**.

13 bits	3 bits	16 bits
Operand address	*Op code*	*Unassigned*
0 12	13 15	16 31

Least sig.   *most sig.*

Layout of an SSEM instruction

Only seven instructions (op codes) were available. They are listed in **the table below**. The upper-case letters in the second column there stand for the following.

- S indicates the currently addressed location in the store.
- A refers to the accumulator.
- C refers to Control (the program counter).

Lower-case letters refer to the contents of the corresponding address (or unit).

Op code	Original notation	Modern description
0	s to C	Absolute indirect unconditional jump
1	s + c to C	Relative indirect unconditional jump
2	- s to A	Load negative
3	a to S	Store accumulator
4	a - s to A	Subtract
5	-	Not used (treated same as subtract)
6	Test	Skip next instruction if accumulator is negative
7	Stop	Halt

SSEM op codes

THE CAMBRIDGE EDSAC

The layout of a 17-bit EDSAC instruction in 1949 was as shown in **the figure below**, and the meanings of the fields were:

5 bits	1	10 bits	1
Op code	**U**	**Address**	**D**
16	12 11	10	1 0

Most sig.   *least sig.*

Key
Op code: a five-bit instruction code. This was designed to be represented by a mnemonic letter (see below), so that for example the ADD instruction used the EDSAC teleprinter's bit-pattern for the letter A; **U**: a spare (unused) bit; **Address**: ten bits for a memory address (also used for other purposes – see below); **D**: one bit to signify whether the instruction operated on a number contained in one word (17 bits) or two words (35 bits).

Layout of an EDSAC instruction

Notice that an EDSAC instruction as displayed on the operator's monitor tubes appeared conventionally, with the least-significant digit at the right-hand end. This is in contrast to the form of display adopted at Manchester and NPL (and for DEUCE), where the display was aligned with the way an engineer viewed serial digits in time sequence on an oscilloscope.

In 1949 the EDSAC instruction set had the following 18 operations.

A n: Add the number in storage location n into the accumulator

S n: Subtract the number in storage location n from the accumulator

H n: Copy the number in storage location n into the multiplier register

V n: Multiply the number in storage location n by the number in the multiplier register and add the product into the accumulator

N n: Multiply the number in storage location n by the number in the multiplier register and subtract the product from the accumulator

T n: Transfer the contents of the accumulator to storage location n and clear the accumulator

U n: Transfer the contents of the accumulator to storage location n and do not clear the accumulator

C n: Collate [logical AND] the number in storage location n with the number in the multiplier register and add the result into the accumulator

R 2^{n-2}: Shift the number in the accumulator n places to the right

L 2^{n-2}: Shift the number in the accumulator n places to the left

E n: If the sign of the accumulator is positive or zero, jump to location n; otherwise proceed serially

G n: If the sign of the accumulator is negative, jump to location n; otherwise proceed serially

I n: Read the next character from paper tape and store it as the least significant 5 bits of location n

O n: Print the character represented by the most significant 5 bits of storage location n

F *n*: Read the last character that was output, so that it may be verified by program

X: No operation

Y: Round the number in the accumulator to 34 bits

Z: Stop the machine and ring the warning bell

EDSAC programmers always prepared their programs in symbolic form with decimal addresses. As an example, suppose that we wish to add together the short (17-bit) values held in storage locations 24 and 73, placing the result in location 66. Suppose that address 41 may be used as a temporary dump. Here is the EDSAC program as it would be punched on to paper tape by the teleprinter and perforator (of course, the comments would not have been punched). The abbreviation 'acc' stands for EDSAC's accumulator register.

Instruction	Comments
T41 S	transfer the contents of acc to address 41 & clear the acc
A24 S	add into acc the contents of location 24
A73 S	add into acc the contents of location 73
U66 S	unload acc into location 66

THE FERRANTI MARK I'S INSTRUCTION FORMAT

From the programmers' viewpoint, the Ferranti Mark I included the following central registers:

- an 80-bit double-length accumulator;
- a 40-bit multiplicand register;
- eight 20-bit index or address-modification registers (called 'B lines');
- a 10-bit program counter.

The instruction set made provision for up to 64 commands, though only 47 of these op codes were actually assigned. The layout of an instruction was as shown in **the figure below**.

10 bits	3 bits	1	6 bits
Operand address, n	B	Spare	Op code
0 9	10 12	13	14 19

Least sig. — *most sig.*

Layout of a Ferranti Mark I instruction

Apart from the usual logical and signed arithmetic operations, there were unsigned versions of several arithmetic functions. Amongst the more unusual commands were:

- sideways add of the bits in a word ('population count');

- giving the position of the most-significant '1' in a word;

- a hardware 20-bit random-number generator;

- sending a pulse to the console's audio amplifier.

A light-hearted use of this last instruction was to play 'computer music'.

Ferranti Mark I programmers split each word into groups of five digits, each of which could be represented by a character from a modified 5-bit teleprinter code. Programmers were expected to memorise this code and to write programs in it. There was no symbolic assembler. Numbers were written with the least-significant position at the left-hand end. By way of illustration, the decimal values 0, 1, 2, 3, 4, 5, 6 and 7 would be written as teleprinter symbols /, E, @, A, :, S, I and U according to the following code.

```
00000    /
10000    E
01000    @
11000    A
00100    :
10100    S
01100    I
11100    U
... etc.
```

For example, the op code /A causes the contents of the accumulator to be stored, and the code /F is 'multiply and add'; both instructions being unmodified (i.e., using B0). If modifier register B7 was used with these two instructions, they would be written as UA and UF. Below is a short program that places in address /C the scalar product of two 18-element vectors whose fixed-point values are stored in the following addresses (inclusive):

Vector (i) in lines /N, @N, ..., LN
Vector (ii) in lines /F, @F, ..., LF

Address	Machine code	Explanation
//	L///	Number of elements in the vectors
/E	IST/	Entry point; set round-off
/@	//Q0	Set B7
/A	/NUK	Add product
/:	/FUF	… to partial sum
/S	A:QG	B7 := B7 − 1 (ie adjust counter)
/ſ	A:/T	Test for last cycle, jumping back to line A if more to do
/U	/C/A	Transfer result to address /C

INSTRUCTION FORMAT FOR THE ENGLISH ELECTRIC DEUCE

The word length of the DEUCE was 32 bits. When used as an instruction, the format and the meanings of the fields are as shown in **the figure below.**

1	3	5 bits	5 bits	2	5 bits	4	5 bits	1	1
U	NIS	Source	Destin	Ch	Wait	U	Timing	U	Go
0	1 3	4 8	9 13	16 20	21 24	25 29	30	31

Least sig   *most sig*

Key
U: unassigned; **N NIS**: Next Instruction Source: indicates long delay line D1 -> D8; **S Source**: number of selected source (short or long delay line); **D Destination**: number of selected destination (short or long delay line); **C Characteristic**: gives length of transfer (0, 1 or 2 -> single, long or double) & drum read/write mode; **W Wait number**: gives first minor cycle* of transfer; **T Timing number**: gives minor cycle* of next instruction, & sometimes also last minor cycle* of transfer; **G Go digit**: If G = 0, wait for handkey to be pressed; if G = 1, full speed ahead.

* a minor cycle is equivalent to a word, ie to 32 digit-periods. W and T were specified in terms of minor cycles. A digit-period was one microsecond.

Layout of a DEUCE instruction

A minor cycle is equivalent to a word, i.e. to 32 digit-periods. W and T were specified in terms of minor cycles. A digit period was one microsecond.

The written short form of an instruction is: < N S D C W T G>. If the current instruction was at position m in a long delay line, then the next instruction is taken from position $(m + 2 + T)$ in the delay line specified by N. If the current instruction was at position m in a long delay line, then the transfer of operands commences at time $(m + 2 + W)$.

An operand-address number for S or D could be in the range 0 to 31. Briefly, values 1 to 12 signify long (32-word) delay lines; values 13 to 21 refer to somewhat shorter lines (see below) and values 22 to 32 are used for special purposes. Specifically:

13, 14, 15, 16: one-word temporary stores referred to as TS13, TS14, TS15, TS16

17, 18: quad-word stores referred to as QS17 and QS18

19, 20, 21: double-word stores referred to as DS19, DS20 and DS21

Certain values of S and D are used to specify operations, or functions. For example, destinations D = 22 and D = 23 signify respectively the additive and subtractive inputs to DS21. Destinations D = 25 and D = 26 signify respectively the additive and subtractive inputs to TS13. Logical operations on TS14 and TS15 are given by S25 and S26. If D = 24, then various values of S in the range 0–11 set or reset various flags and triggers. One such trigger starts the multiplication process, which could then be overlapped with other operations. Another trigger is used to pulse the console's audio amplifier (hooter). If D = 27 or D = 28, the 'source' operand is tested according to one of two conditions and, depending upon the result of the test, the instruction from one minor cycle later than the normal instruction is obeyed. This therefore gives conditional branching – (called 'discrimination' in the original documentation). Input–output is available via S = 0 and D = 29. Transfers to or from the magnetic drum are controlled by D = 30 and D = 31.

In conclusion, the meaning of different combinations of the S and D values was quite complex. By choosing appropriate combinations, the equivalent of about 30 effectively distinct operations could be obtained. By carefully choosing values of the N, W and T fields of each instruction, the execution time of a program could be minimised – this process on the part of a skilled user giving rise to the term 'optimum programming'.

APPENDIX B
TURING AND COMPUTING: A TIMELINE

Below is a chronology of computer-related events in the period that Alan Turing (AMT) spent at NPL and at Manchester, 1945–54. This timeline has mostly been compiled from information in the first edition of Andrew Hodges' excellent biography *Alan Turing: the Enigma*, published by Burnett Books in 1983. Supplementary data on the Pilot ACE project has been taken from *Turing's Legacy: a history of computing at the National Physical Laboratory, 1945–1995*, by D M Yates, published by the Science Museum in 1997. Finally, additional information on Turing's time at Manchester comes from personal contact with the survivors of that period.

ALAN TURING AT NPL, 1945–8

1945: June   J R Womersley meets AMT and arranges for him to join the Mathematics Division at NPL.

1945: September   AMT and Don Bailey experiment (unsuccessfully) with an air-filled acoustic delay line at Hanslope Park.

1945: 1 October   AMT takes up the post of Temporary Senior Scientific Officer at NPL. He is placed on his own, in a special section tasked with designing an electronic universal computing machine. This is soon referred to as ACE (automatic computing engine).

1945: End December   AMT's ACE Report is completed.

1946: Spring   Womersley writes an enthusiastic memorandum on AMT's ACE Report, for tabling at NPL's Executive Committee meeting of 19 February. Full discussion postponed until the 19 March meeting, at which AMT is invited to present his ideas. Also present is Sir George Nelson, the Director of English Electric. The ACE proposal is approved in principle, and on 17 April Sir Charles Darwin, the NPL Director, requests £10,000 to be allocated for a Pilot version of ACE. The Department of Scientific and Industrial Research (DSIR, the parent body for NPL) agrees to this expenditure on 8 May.

1946: May   AMT is allocated two Scientific Officers: J H Wilkinson (half-time) and Mike Woodger. AMT continues to refine his design, which, by May, has reached Version 5. At this stage the project lacks any formal electronic engineering support from NPL. (Dollis Hill began some modest work for NPL on mercury delay lines in April, but their effort is limited by more urgent GPO priorities.)

1946: August   Darwin approaches TRE, to see whether F C Williams could help NPL with CRT storage technology for the ACE project.

1946: 22 November   Deputation from TRE visits NPL to discuss the ACE project. TRE reports that 'although an elaborate paper design [for ACE] has been laid down, the fundamental problem of storage of information has not yet been solved'. According to A M Uttley (TRE), it is at this meeting that AMT and F C Williams appear to disagree strongly over the best way to design high-frequency pulse circuits. By mid December Williams has taken up an appointment as Professor of Electrical Engineering at the University of Manchester. No help from TRE for ACE is forthcoming.

1946–7: November–April   Darwin tries, but fails, to get engineering support for the ACE project from the universities of Manchester (F C Williams) and/or Cambridge (M V Wilkes) – each of which is by then independently engaged on designing and building its own stored-program computer. Meanwhile, during this period AMT gives several newspaper and radio interviews about his ideas for universal automatic computers.

1946–7: December–January   NPL arranges that AMT should give a course of lectures in London on computers, to about 25 invited electronic engineers and similar people. In the event, J H Wilkinson takes over the lecturing because –

1946: 26 December   – at the suggestion of Darwin, AMT sets sail for America to be the only British attendee at the Symposium on Large-Scale Calculating Machinery held at Harvard from 7 to 10 January 1947. After the Symposium AMT spends two weeks at Princeton. In his trip report, AMT later comments that 'the Princeton group [of computer designers] seem to me much the most clear headed and far sighted of these American organisations, and I shall try to keep in touch with them'.

1947: February   At the suggestion of Hartree, H D Huskey (a member of the American ENIAC team with some engineering experience) arrives to spend a year at NPL.

1947: Spring   Huskey, anxious to make a start on building a computer, proposes 'Version H', a simplification of AMT's Version 5 of ACE. Huskey, Wilkinson and Woodger plan an electronic implementation of Version H, also known as the 'Test Assembly'. AMT boycotts these developments and starts to put together a few circuits of his own, in the cellar of the Mathematics Division's building at NPL.

1947: Summer   Darwin sets up a special Electronics Section under H A Thomas, in the existing Radio Division at NPL. However, Thomas does not immediately appreciate what is needed in order to move the ACE project forward. AMT loses interest in the ACE project.

1947: 23 July   AMT is granted a sabbatical year at Cambridge, on half-pay from NPL.

1947: 18 August   At a special meeting, this day is declared as the official start to the building of the Pilot version of ACE. AMT attends the meeting but says nothing.

1947: September   NPL appoints E A Newman and D O Clayden, two experienced electronic engineers who have worked for EMI and for the legendary A D Blumlein on radar during the war. They join the Electronics Section to work on the Pilot ACE project.

1947: 30 September   AMT returns to Cambridge to resume his King's Fellowship, after a break of nearly eight years. The Fellowship is set to run until March 1952.

1947: November   Work on Huskey's Test Assembly stopped by Darwin, because of complaints by Thomas (of the Electronics Section). Huskey, Wilkinson and Woodger reduced to writing a report on numerical analysis and programming issues; this report appears in April 1948.

1947–8: Winter–spring   AMT finishes a numerical analysis paper on the ACE work – 'Rounding-off errors in matrix processes', published in the *Quarterly J Mech Applied Maths,* vol 1, 1948. He becomes more interested in thinking processes and mechanised learning and renews his interest in game theory. J H Wilkinson visits AMT at Cambridge from time to time but can report no real progress with the implementation of the Pilot ACE.

1948: April   NPL makes the Electronics Section a separate unit, led by F M Colebrook, who immediately takes the ACE project in a positive direction. Critically, he arranges that Wilkinson,

Woodger, G G Alway and D W Davies from the Mathematics Division should temporarily join the Electronics Section to work with Newman and Clayden on implementing the Pilot model of ACE. English Electric seconds two engineers and four technicians to the project. In the event, Pilot ACE will not run its first program until 10 May 1950.

1948: May   AMT visits NPL and sees no point in returning there after his sabbatical.

1948: 28 May   AMT accepts an offer of employment from Manchester University and gives notice of his resignation from NPL but retains Cambridge as his base for the summer. The Manchester appointment is to run from 29 September, the post being entitled 'Deputy Director of the Computing Machine Laboratory' with the status of a Reader in the Mathematics Department under Professor Newman. His salary is the first call upon a Royal Society grant awarded earlier to Newman. The grant, approved in July 1946, is at the rate of '£3,000 a year for five years for salaries, together with the sum of £20,000 to be spent on construction during the same period'. AMT's salary is £1,200 p.a. (increased to £1,400 p.a. in June 1949). In the event, the 'construction' money will later be used to provide a new building – the Computing Machine Laboratory – to house the Ferranti Mark I computer in 1951.

1948: July–August   AMT completes a lengthy report for NPL on 'Intelligent Machinery'. In September 1948 Darwin judges the report 'not suitable for publication' and it is filed away at NPL. An edition of this paper is, much later, published in *Machine Intelligence 5*, Edinburgh University Press, 1969.

1948: 8 July   AMT receives a reply from F C Williams to an enquiry about the instruction set for the Manchester computer (the Small-Scale Experimental Machine – SSEM, sometimes called 'the Baby' – has run a program on 21 June). AMT writes a long division routine for the SSEM and posts this to Manchester. He sends another factoring routine off on 2 August before going away on holiday to Switzerland, followed by some time in the Lake District and then a third holiday in Wales.

ALAN TURING AT MANCHESTER, 1948–54

1948: October   AMT moves to Manchester shortly after 2 October. Remains a Fellow of King's College, Cambridge, regularly spending August back at Cambridge. At Manchester, he chooses to live in a large lodging house in Hale, Cheshire, outside the conurbation.

1948: October   On seeing the SSEM at close quarters, AMT advises F C Williams's team that 5-track Creed paper tape equipment should be connected to the computer for input–output (thus replacing the existing manual I/O). AMT arranges for such equipment to be 'obtained' (via his GCHQ or GPO contacts?). A research student, Dai Edwards, is given the task of connecting this equipment to the SSEM.

1948: 26 October   Sir Ben Lockspeiser, the Government Chief Scientist, places an order with Ferranti Ltd to construct an electronic calculating machine 'to the instructions of Professor F C Williams'. This was to become the Ferranti Mark I, the commercial version of the Williams–Kilburn research project at the university.

1948–9: Autumn–spring   The SSEM's hardware is under extensive development and enhancement. By now called the Manchester University Mark I, or sometimes MADM, the computer working in April 1949 includes a drum backing store and two 'B lines' (index registers). AMT's name is not mentioned on any of the 34 Manchester Mark I patents registered in the period 1946–9. Max Newman's name is on one patent (the B line patent) along with those of F C Williams, T Kilburn and G C Tootill. AMT's role in the Manchester Mark I design seems primarily to be the specification of the input–output instructions, devising a programming system and writing the bootstrap routine.

1949: June   Newman specifies a test problem for the Manchester Mark I: investigating Mersenne Primes. Kilburn and Tootill code up a program for this and AMT subsequently writes a faster version.

1949: Summer   AMT (it is believed) works largely at home, rather than at the university. Perhaps he is developing his ideas about machine intelligence. Nevertheless, in a paper presented in Cambridge at the Conference on High-Speed Automatic Calculating Machines (22–5 June), he considers the problem of program correctness.

1949: August   AMT writes a paper on 'The word problem in semi-groups with cancellation'. This is later published in *Ann. Math.* (Princeton) 52, 1950.

1949: Autumn   By October an enhanced version of the Williams–Kilburn computer has been developed, which includes program control of drum transfers and programmed input–output.

The input–output routines, using 5-bit teleprinter code, have most probably been written by AMT. This version of the Manchester Mark I is to perform useful work in the next nine or so months, including investigation of the Riemann hypothesis (Zeta function) and calculations in optics (ray tracing).

1949: October   AMT takes on a research student, Audrey Bates, who has graduated from Manchester in July with a first in mathematics. At about the same time Cicely Popplewell, a Cambridge mathematics graduate 'with experience of punched cards used in housing statistics', is employed by Manchester University to help generally with system programming for the prototype computer. These two women share AMT's office, but he remains relatively uncommunicative to both.

1949: 27 October   A formal Discussion on 'The Mind and the Computing Machine' is held in the Philosophy Department at Manchester University. AMT writes up his views and submits a paper to the philosophical journal *Mind*. This 27-page paper (appearing in 1950) poses the question 'Can machines think?' The well-known Turing Test arises from this work.

1949: November   Computer specifications are being passed from the university to Ferranti Ltd at Moston, near Manchester. AMT undoubtedly takes part in this process. He writes Appendix 2, 'Generation of random numbers', for the document entitled *Informal report on the design of the Ferranti Mark I computing machine* by G C Tootill, 22 November 1949.

1950: 8 February   AMT writes that he is working on his 'mathematical theory of embryology' (morphogenesis), with the aim of addressing five problems: (a) gastriculation; (b) polygonally symmetrical structures, e.g. starfish or flowers; (c) leaf arrangements, in particular the way the Fibonacci series comes to be involved; (d) colour patterns on animals, e.g. stripes, spots and dappling; (e) pattern on nearly spherical structures such as radiolaria, 'but this is more difficult and doubtful'. He thinks that this work 'is not altogether unconnected with' his interest in brain cells and the physiological basis of memory and pattern recognition.

1950: June   Thanks to an unusually long error-free run (3 p.m. through to 8 a.m. the next day) AMT uses the Manchester Mark I to investigate the Zeta function. The results are published as 'Some calculations of the Riemann Zeta function', *Proc. London Math. Soc.*, vol. 3, 3, 1953. At about this time he also writes a program that demonstrates unpredictability.

1950: Summer   AMT buys a Victorian semi-detached house in Wilmslow, an outer suburb of Manchester. Decides not to install a telephone.

1950: October   Audrey Bates submits an MSc thesis entitled *On the mechanical solution of a problem in Church's lambda calculus.* The degree is to be awarded in January 1951, by which time she has left the university to join Ferranti Ltd as a programmer.

1951: 12 February   The Ferranti Mark I computer is delivered to the university. After installation and testing, it is, however, not giving acceptable error-free service until about June.

1951: March   AMT produces the *Programmers' Handbook* for the computer.

1951: April   AMT revisits Group Theory, particularly the 'word problem for groups'. He derives a result that J H C Whitehead (Oxford) finds 'sensational'. The result is not published.

1951: 9–12 July   Inaugural Conference for the Manchester University Computer (i.e., the production version, the Ferranti Mark I). AMT presents a paper entitled 'Local Programming Methods and Conventions'. From this point onwards, his main research topic appears to be morphogenesis.

1951: July   Christopher Strachey visits the Computing Machine Lab at Manchester for the first time. He intends to write a draughts program but is persuaded by AMT to write an interpretive trace program, roughly equivalent, in effect, to the machine simulating itself. Some weeks later Strachey returns to test this program and, to the amazement of all, gets it right in a very short time. This is by far the largest (1,000 machine instructions) program that has ever been attempted on the Ferranti Mark I up to this time. As a result, Turing recommends Strachey to NRDC. Lord Halsbury interviews Strachey in November 1951, and Strachey joins NRDC formally in June 1952.

1951: October   R A Brooker arrives from Cambridge and takes over from AMT the responsibility for software development and systems organisation in Manchester.

1951: November   AMT completes his main paper on morphogenesis, incorporating his mathematical theory of embryology. He regards the importance of this paper as 'the equal of "Computable Numbers"'. The paper is published in August 1952 as 'The chemical basis of morphogenesis', *Phil. Trans. Royal Soc.* B237. AMT is to leave much more morphogenesis work unpublished.

1952: January   By this month Alick Glennie, employed by the Atomic Weapons Research Establishment (AWRE), has begun to book time on the Ferranti Mark I computer at Manchester. He is one of a number of outside users from government, industry and academia who come to use the Manchester facilities.

1952: 23 January   AMT's house in Wilmslow is burgled.

1952: 11 February   AMT arrested on a charge of Gross Indecency and released on bail.

1952: 12 February   AMT gives a talk on morphogenesis to the Ratio Club in London.

1952: 26 February   AMT appears at the Wilmslow Magistrates' Court.

1952: 29 February   AMT completes the revisions to his paper on morphogenesis.

1952: 15 March   AMT completes a paper on the Zeta function, resulting from his earlier calculations on the Manchester computer.

1952: 31 March   AMT appears at the Quarter Sessions in Knutsford; bound over for a year and obliged to undergo organo-therapy 'treatment'.

1952: Spring–summer   Strachey comes for short periods to Manchester and writes the 'Love Letters' and Draughts programs, the latter being completed and written up by the start of September.

1952: Summer   Alick Glennie (from AWRE) uses the Manchester computer for atomic weapons calculations and develops his private 'autocode' system. AMT and Glennie simulate playing chess with the computer, AMT reproducing on paper the moves that his chess-playing algorithm would have made if it had been coded up and run on the machine.

1952: Autumn   AMT continues experimenting on the computer with solutions to the difficult differential equations arising out of the chemical theory of morphogenesis. His occasional advisory work for GCHQ on cryptanalysis, which may have been ongoing since his Bletchley Park days, probably ceased at this time.

1953: 15 May   AMT is appointed to the specially created position of Reader in the Theory of Computing at the University of Manchester, with effect from 29 September, when his existing five-year appointment is due to run out.

1953: Summer   AMT takes on a second research student, Bernard Richards, who works on AMT's theories of morphogenesis, amongst other things solving one of his equations and showing that this can accommodate a few of the simpler patterns found in monocellular radiolaria.

1953: Autumn   Tom Kilburn's engineering group at Manchester produces a small transistor computer, which first runs a program in November. AMT plays no part in this project. AMT continues to work on morphogenesis, also spending some time on the Theory of Types with Robin Gandy.

1954: March   Tony Brooker releases the Mark I Autocode, regarded by many as the first publicly available high-level programming language; AMT is reportedly not interested.

1954: May   Meg, the successor computer to the university's Mark I, first runs a program. Meg has been designed by Tom Kilburn's group. AMT plays no part in this project.

1954: 7 June   Alan Turing kills himself, being found in his house at Wilmslow having apparently eaten an apple dipped in cyanide. His death is completely unexpected, coming as a great shock to all who know him.

APPENDIX C
FURTHER READING

Anyone interested in the life of Alan Turing should start with Andrew Hodges' classic 600-page biography: *Alan Turing: the Enigma*, published by Burnett Books in 1983 (ISBN: 0-09-152130-0). Unfortunately, this otherwise very carefully researched book does not give much detailed information on Turing's computer design activities.

For illustrated simple explanations of the terminology, technology and programming of early computers, see the book *Early British Computers* by Simon Lavington. This is out of print but has helpfully been made available at http://ed-thelen.org/comp-hist/EarlyBritish.html

Alan Turing did not publish any specific paper on computer design, his work in this area being confined to internal reports. Fortunately, the more important of these reports are reproduced in a book edited by B J Copeland, called *Alan Turing's Automatic Computing Engine*, published by Oxford University Press in 2005 (ISBN: 0–19–856593–3).

The original scientific papers describing most early computing activity in Britain were published in specialist journals such as the *Proceedings of the Institution of Electrical Engineers* and, from 1956 onwards, in the *Journal of the British Computer Society*. From 1979 onwards, retrospective histories started to appear in a new international journal called the *Annals of the History of Computing*. As far as the material in this book is concerned, Alan Turing's three most relevant original papers are:

- 'On Computable Numbers, with an application to the *Entscheidungsproblem*', *Proceedings of the London Mathematical Society*, series 2, vol. 42, 1936–7;

- 'Computing machinery and intelligence', *Mind* vol. 59, 1950, pages 433–60;

- 'The chemical basis of morphogenesis', *Philosophical Transactions of the Royal Society*, B237, August 1952.

These papers are helpfully reproduced in another book edited by B J Copeland, entitled *The essential Turing: the ideas that gave birth to the computer age*, and published by Oxford University Press in 2004 (ISBN: 978–0-19–825079–1).

Since the original specialist papers covering the early British computers are difficult for the general reader to obtain, the technical details of the more important commercially available machines have been made available at the *Our Computer Heritage* website, organised by the Computer Conservation Society – see www.ourcomputerheritage.org/

Finally, here is a list of useful books that give (retrospective) accounts of many of the early computing projects. The interested reader may consult these references and those listed above to find the sources for all quotations in this book.

GENERAL ACCOUNTS OF THE PERIOD 1945–60

Bowden, B V (ed) (1953) *Faster than thought.* Pitman Press, London.
Metropolis, N, Howlett, J and Rota, G-C (eds) (1980) *A history of computing in the twentieth century.* Academic Press. ISBN: 0–12–491650–3.
Campbell-Kelly, M and Aspray, W (1996; 2nd edition 2004) *Computer: a history of the information machine.* Basic Books. ISBN: 0–465–02989–2.
Lee, J A N (1995) *Computer pioneers.* IEEE Computer Society Press. ISBN: 0–8186–6357–X.
See also the web version of the book *Early British Computers* at http://ed-thelen.org/comp-hist/EarlyBritish.html

CHAPTER-SPECIFIC BOOKS

CHAPTER 2
Yates, D (1997) *Turing's Legacy: a history of computing at the National Physical Laboratory, 1945–1995.* Science Museum, London. ISBN: 0–910805–94–7.

CHAPTER 3
Wilkes, M V (1985) *Memoirs of a computer pioneer.* MIT Press. ISBN: 0–262–23122–0.
Ferry, G (2003) *A computer called LEO: Lyons teashops and the world's first office computer.* Fourth Estate. ISBN: 1–84115–185–8.
See also the Cambridge history site: www.cl.cam.ac.uk/conference/ EDSAC99/

CHAPTER 4

Lavington, S (1998) *A history of Manchester computers* (2nd edition). BCS. ISBN: 0–902505–01–8.

See also the Manchester history site: www.computer50.org/

CHAPTER 5

Lavington, S (2011) *Moving targets: Elliott-Automation and the dawn of the computer age in Britain, 1947–67.* Springer. ISBN: 978–1–84882–932–9.

CHAPTERS 6 AND 7

Campbell-Kelly, M (1989) *ICL – a business and technical history.* Oxford University Press. ISBN: 0–19–853918–5.

INDEX

Locators in *italics* indicate photographs

109

Lightning Source UK Ltd.
Milton Keynes UK
UKOW020949131112

202103UK00003B/56/P